EDUCATION AND TRAINING
FOR CLINICAL CHEMISTRY

EDUCATION AND TRAINING FOR CLINICAL CHEMISTRY

Prepared for Publication by

MARTIN RUBIN, PhD,
Georgetown University Medical Center,
Washington, DC, USA

PER LOUS, MD, PhD,
Bispebjerg Hospital,
Copenhagen, Denmark

Published for The International Federation of Clinical Chemistry
Committee on Education and Training in Clinical Chemistry
by MTP Press Limited

Published in the UK
for IFCC by
MTP Press Limited
St Leonard's House,
Lancaster,
England

Copyright © 1977 International Federation of Clinical Chemistry
Softcover reprint of the hardcover 1st edition 1977

First published 1977

ISBN-13: 978-94-011-6178-7   e-ISBN-13: 978-94-011-6176-3
DOI: 10.1007/978-94-011-6176-3

A. Wheaton & Co., Exeter

Contents

Individual Countries:

Preface

While the historic roots of clinical chemistry originate from the chemical sciences the growth of the subject has been dependent upon the political, social, economic and technologic national soil in which it has developed. Thus the present leaders in this field have backgrounds variously in chemistry, medicine, pharmacy or sometimes biology. Today, clinical chemistry has attained stature as a unified independent discipline. It is characterized by active and productive international and national societies; its function codified in the law of many countries; its scientific content the sole subject of international and national journals as well as textbooks and educational programs; and its international, regional and national meetings have become focal points for major exchange of scientific, clinical and technical information and exhibition. The positive impact of the discipline upon the delivery of health care has given it a significant position in the economics of public health. As a consequence it has become the most rapidly-growing segment of the industrial and commercial component of health maintenance.

These changes have brought the need to define the educational and training processes to prepare future leaders of clinical chemistry. The diverse backgrounds of the present directors of clinical chemical laboratories has required that the viewpoints of chemists, pharmacists, physicians and biologists be brought into harmony. This has been achieved by the years of discussion, debate and review by colleagues of varied professional backgrounds. This monograph reflects their consensus viewpoint for the practice of clinical chemistry at its most advanced level.

As an essential part of this task it was important to gather as much information as possible about the status of clinical chemistry throughout the world. The results of this survey conducted mostly between 1972 and 1976 is incorporated in the last section of the text. The rapidly-changing nature of clinical chemistry in all the countries of the world means, inevitably, that the printed information is already dated. Errors, incomplete reporting and all the other inaccuracies of this kind of document are the responsibility of the editors.

Our colleagues have been patient and invariably most helpful in the years in which this effort has gone forward.

The members of the International Union of Pure and Applied Chemistry, Section of Clinical Chemistry, Commission on Teaching of Clinical Chemistry have been: M. Rubin, PhD, Chairman; P. Lous, MD, Secretary; D. Curnow, PhD; A. Latner, MD; A. Defalque, D.Pharm.; Pharm.; J. Porter, PhD; and M. K. Schwartz, PhD.

The contributors to the Committee on Education of the International

Federation of Clinical Chemistry have been: R. Balado, F. Fares Taie (Argentina); D. H. Curnow, W. Roman (Australia); E. Kaiser (Austria); W. Blomme, A. Defalque, A. DeLeenheer, J. P. Dinant, J. Duvivier, R. Ruyseen, E. Tytgat, H. Wachsmuth (Belgium); M. Mostajo Baya (Bolivia); D. M. Nogueira, G. Hoxter (Brazil); Y. T. Todorov (Bulgaria); D. J. Campbell, S. H. Jackson, R. H. Pearce, J. Porter (Canada); J. Lizana (Chile); J. Homolka, J. Horejsi, V. Hule, K. Masek (Czechoslovakia); R. J. Haschen, H. J. Raderecht (DDR); P. Astrup, O. Lauritsen, P. Lous (Denmark); R. Chediak (Ecuador); M. M. Abdel Kadar (Egypt); K. B. Bjornesjo (Ethiopia); H. Breuer, J. Büttner, G. Hillman (Federal Republic of Germany); H. Adlercreutz, N. Saris (Finland); J. E. Courtois, L. Hartman, P. Louisot, P. Metais (France); D. A. Mensah, G. R. E. Swaniker (Ghana); A. Fischer, I. Horvatth, B. Ringelhann, J. Sos (Hungary); A. Baghdiantz, H. Gauguik (Iran); S. Rangaswami (India); T. G. Brady, R. Cahill, W. C. Love (Ireland); C. Castelli, I. Masi, A. Rossi, G. Vanzetti (Italy); Z. Tamura, Y. Yamamura (Japan); R. Perez Herrera, M. L. Castillo de Sanchez (Mexico); Th. Strengers, E. J. van Kampen (Netherlands); B. K. Adadevoh, P. A. Akinyanju (Nigeria); L. Eldjarn, H. Palmer (Norway); J. Krawczynski (Poland); G. da Costa, J. A. Lopez do Rosario (Portugal); S. Comorosan (Rumania); L. S. de Villiers (South Africa); C. H. de Verdier (Sweden); H. Ch. Curtius, M. Roth (Switzerland); J. R. Daly, H. Lehmann, F. Mitchell, H. G. Sammons, J. H. Wilkinson (United Kingdom); P. Besch, D. Birenbaum, B. E. Copeland, H. D. Gruemer, R. M. Hackman, D. R. Harms, E. W. Hull, A. Kaplan, D. A. H. Roethel, A. M. Salton, M. K. Schwartz (United States of America); E. G. Larsky, V. V. Menshikov, V. N. Orekhovich (USSR); G. Gonzales (Venezuela); M. Mikac Devic, I. Ruzdić, B. Straus (Yugoslavia).

1

Background

The problems of education and training in clinical chemistry were of early concern when the Commission of Clinical Chemistry of the Division of Biological Chemistry of the International Union of Pure and Applied Chemistry (IUPAC) and the International Federation of Clinical Chemistry (IFCC) were jointly and simultaneously founded in 1952. Interest in these questions, manifest in the organizational charters and statement of objectives of both organizations was enhanced by the ratification of the IFCC constitution and by-laws by eighteen national societies of clinical chemistry in 1962. The subject became a matter for agenda discussion at the joint IUPAC and IFCC meeting in Paris in 1963. At that time the IUPAC Commission established a subcommittee on Teaching of Clinical Chemistry. The start of the present effort can be traced to this body which was the first to be entirely devoted to this subject. With the administrative and organizational separation of the IUPAC Commission of Clinical Chemistry and the IFCC, foreshadowed at the VI International Congress of Clinical Chemistry in Munich in 1966 and consummated at the IUPAC meetings in Prague in 1967, the basis was set for a substantive effort on the part of both organizations.

Upon the abolition of the IUPAC Division of Biochemistry in 1967, the former Commission of Clinical Chemistry was elevated to section status. The newly created section established the present IUPAC Commission on Teaching of Clinical Chemistry in 1967. At the same time the IFCC appointed a Committee on Education and Training in Clinical Chemistry to collaborate with the IUPAC Commission. The relations of the two groups and the nature of their respective and joint efforts have provided a good example of the roles and functions of the IUPAC and IFCC organizations in relation to the worldwide developments of clinical chemistry. On the one hand IUPAC provides a forum for the formulation of major concepts whose definition and implementation can be accommodated by the national societies which comprise the International Federation of Clinical Chemistry. The result of this iterative interaction should produce a sound theoretical and practical base for the continued growth of the discipline of clinical chemistry.

THE IUPAC COMMISSION ON THE TEACHING OF CLINICAL CHEMISTRY

The objectives of the Commission and its task of advancing the teaching of clinical chemistry by appropriate activities were set forth with the formation of the Commission in Prague in August 1967. In an exchange of views the members of the Commission decided to undertake a survey of the status of clinical chemistry throughout the world, to review the potential developments in the field as they affected professional requirements, and to propose guidelines for future development. In consonance with the nature of IUPAC, the Commission was organized to serve as a body of international experts without regard to specific national interests. The Commission met 12 times between 1968 and 1976, three of these meetings being held jointly with the members of the IFCC Committee.

The Commission assumed responsibility for the planning, orgnization and drafting of this monograph. Early in its deliberations it was recognized that important suggestions for educational programs in clinical chemistry would be forthcoming from those with a primary background in medicine, the sciences or pharmacy. Consequently subcommittees were appointed with members in these fields. Their thoughts and suggestions were woven into the fabric of this report. At all stages consultation and exchange of views were fostered with the IFCC Committee. In the intervals between Commission and Committee meetings, the work proceeded by intensive correspondence. When a penultimate draft of the document was available in late 1973, the IUPAC elected to publish only a portion of the effort as an Information Bulletin relinquishing the responsibility to IFCC for publication of the full text.

THE IFCC COMMITTEE ON EDUCATION AND TRAINING IN CLINICAL CHEMISTRY

The IFCC Committee was established in Prague in August, 1967. The stated objectives of the Committee, in keeping with the constitution of the IFCC were to 'advance the science and practice of clinical chemistry and to enhance its services to health and medicine through education and training'. In contrast to the IUPAC organization, the Federation provides a forum for the presentation and exchange of national views in clinical chemistry. For these reasons it was deemed appropriate that the Federation undertake responsibility for the assembly and organization of information relating to the status of clinical chemistry in the various countries of the world. The analysis of this material, its organization and its interpretation in terms of future developments was agreed to be undertaken as a joint task with the IUPAC Commission on Teaching of Clinical Chemistry.

Whenever possible the official viewpoints of the various National Societies were solicited and incorporated into the document. Where this was not feasible, knowledgeable individuals were requested to prepare a report for their country or region. As stated previously, the Committee met jointly with the IUPAC Commission on three occasions. A continuing exchange of documents by correspondence permitted the attainment of the harmony of viewpoint represented in this monograph.

2

Introduction

With the emergence of clinical chemistry as an independent science it has become possible to define its scope and function. In the broadest sense it applies chemical knowledge to an understanding of the functioning of the human body. Its objective is to provide information to the physician for the preservation of health and the diagnosis, prognosis and treatment of disease. This requires not only the application of fundamental principles of chemistry and medicine but also contributions from other major disciplines such as biology, physics, mathematics, electronics, physiology, pharmacology, and toxicology. Clinical chemistry has a basic role in education for the health professions in health care delivery and in research.

This century has observed a changing attitude to health problems. With increasing clarity it has become evident that health and disease cannot be regarded entirely as personal matters but are also the concern of society as a whole. Governments of varied political outlook have declared that they aim to create living standards and health services which would produce optimal conditions for the attainment of ' . . .complete physical, mental, and social well-being and not merely the absence of disease or infirmity'*. Associated with this philosophy has come the recognition that economic progress cannot readily advance where public health is deficient. The responsibility of national and local governments in combating the major communicable diseases was already established in most countries at the turn of the century and epidemic and infectious diseases have ceased to be a real menace to the populations of many countries. These, unfortunately, represent only a minority of the world's people.

In the industrial countries, cardiovascular diseases, cancer, diabetes, and other diseases of civilization (their distinction is their high incidence where people have a life expectancy of more than 60 years) are of growing importance as health problems.

For the latter group of diseases, chemical laboratory investigations have special and very important roles in diagnosis, in control of treatment and in the elucidation of pathogenesis. In the care of surgical patients the laboratory is also of tremendous importance due to a better understanding of pathophysiology and the development of laboratory methods to assess and monitor the condition of patients. Likewise problems of industrial toxicology and health effects of environmental

* Constitution of WHO, Preamble.

pollution have imposed new responsibilities on the discipline. It is therefore not surprising that the general public, as well as health authorities and governmental bodies in several countries, have expressed an increasing interest in the laboratory and the quantity and quality of its services.

In a later section a short summary is provided of the historic development of clinical chemistry. This occurred as the result of social and economic change, advances in other sciences, major contributions from technology and continued evolution within the discipline. Although, as will be evident subsequently, there are many differences from country to country, the course, the tendencies and the difficulties to be faced are much the same throughout the world. The clinical chemistry laboratory has matured rapidly during the past two or three decades and as a consequence now has an indispensable role in assisting clinicians in every branch of hospital and private practice. In addition the impact of modern technology has provided new possibilities for improved health care delivery from augmented laboratory services. Recognition that patients may respond individually to therapeutic agents has created new and difficult analytical problems to monitor plasma drug concentrations. The application of sophisticated equipment to a widening demand for laboratory investigations represents an increasing cost burden which must be borne as part of the total delivery of health care. Yet despite acknowledged progress, it has been demonstrated in many surveys that the quality of the analytical work may frequently be below acceptable standards.

There is every sign that the progress stemming from today's research will eventually be introduced into routine usage. Schemes for population screening for diseases detectable by clinical chemical tests are now being evaluated and widely discussed. The realization of the progress implicit in these advances will rest in large measure upon the education and training of those engaged in clinical laboratory activity. It is hoped that this monograph will contribute toward this objective.

THE DEVELOPMENT OF CLINICAL CHEMISTRY

Historical development

Clinical chemistry has developed from various roots, some of which date far back into the past. Rather than giving a chronological and complete description of its complex history, the following section reviews some of the major concepts and developments which emerged as milestones in the history of this discipline.

Origin of the term 'clinical chemistry'

The term 'clinical chemistry' is still comparatively young; it was not until after the Second World War that it came into common use. In the nineteenth century, the terms 'pathological' or 'medical' chemistry were used. L. Lichtwitz, a clinician, was the first to use the term 'clinical

4

chemistry' as the title of a textbook in 1918. A milestone in the general acceptance of the term 'clinical chemistry' came with the epoch-making two-volume work by Peters and van Slyke (1931). The contents of this work marked out the entire range of the discipline for the first time.

Chemical analysis of body fluids for the diagnosis of disease
The idea of making use of deviations in the chemical composition of body fluids for the diagnosis of disease was brought forward, probably for the first time in a definitive way, by Robert Boyle about 1680, at the suggestion of his friend, John Locke. It took another 150 years, however, for the basic requirements of chemical analysis to be created especially by the work of J. J. Berzelius, J. von Liebig and their disciples.

About 1840, after a series of isolated findings on the relation of chemical composition of body constituents and disease had become known this concept was taken up again by several investigators (Simon, Andral, Heller, Becquerel, Bence Jones).

Prior to that time, clinical medicine was entirely under the influence of the Romantic School and its speculative natural philosophy. However, after the publication of von Liebig's pioneer monograph, *Die Tierchemie oder Organische Chemie in ihrer Anwendung auf Physiologie und Pathologie,* clinicians such as Schoenlein and von Frerichs, and pathologists such as Virchow, lent support to the use of chemical methods for diagnostic purposes.

In the time that followed, various instructions for the chemical analysis of urine and then of blood were published, making the new diagnostic methods widely known (e.g. Rees, Neubauer). By the end of the nineteenth century, the concept of chemical analysis of body fluids as an important diagnostic aid had gained general acceptance.

Introduction of the concept of homeostasis
Some 100 years ago, under the influence of the new cell theory developed early in the nineteenth century, Claude Bernard established the principle of the constant *milieu interieur* of cells. This concept has subsequently been extremely important to the entire field of biology and medicine. Further developing Bernard's ideas, Walter B. Cannon, in 1926 coined and defined the term 'homeostasis'. This concept has proved very fruitful for clinical chemistry, particularly in recent times. It was useful in the process of replacing the rather more static approach of the early days which focused attention on chemical composition, with the dynamic approach of modern clinical chemistry. This concept became a guiding principle for the experimental research of the last 50 years. In practice, today's clinical chemistry is largely the clinical chemistry of extracellular fluid, i.e. in Bernard's terminology, of the *milieu interieur.*

One of the first and most important applications of this development was the concept of acid-base balance. In the nineteen-twenties through the pioneer work of van Slyke in particular, both the experimental methods and a clinically applicable basis were created for the diagnosis and therapy of disorders in acid-base balance.

Introduction of colorimetry and photometry as analytical tools

In the course of the nineteenth century, quantitative analysis of urine and blood was carried out using the traditional methods of gravimetric and titrimetric analysis. The insufficient sensitivity and the troublesome and time-consuming nature of these methods impeded progress toward a broader use of clinical chemical investigation. Thus, the introduction of colorimetric and photometric methods near the end of the century was a major turning-point. These methods were first applied to the determination of hemoglobin (Gowers 1878, Vierordt 1873). The breakthrough came with the work of O. Folin, however, who developed a large number of photometric methods beginning with the determination of creatinine in 1904. Photometric measuring devices began to appear and Duboscq was the first to develop suitable equipment. Towards the end of the nineteenth century, Pulfrich introduced his 'Stufenphotometer' and L. Heilmeyer, in particular, advocated its wider application to clinical chemistry. In the late twenties, the first photometers with photoelectric cells were described. Today, photometry occupies an outstanding position in advanced analytical clinical chemistry.

Enzyme activities as indicators of cellular and organ disorders

With the measurement of enzyme activities in body fluids, clinical chemistry gained a new resource of great diagnostic importance. The concept of using enzymes, released from cells, as sensitive indicators of cellular or organ disorders, was first realized by Wohlgemuth (1908) when measuring the activity of the alpha-amylase in urine in cases of acute pancreatitis. For the next 50 years, enzymes as diagnostic aids were not developed much further as sensitive methods for the determination of enzyme activities were unavailable. Thus, even by 1950, apart from the alpha-amylase determination, only the analysis of acid and alkaline phosphatases and of lipase were in standard use. Once it had been shown that cellular enzymes entered the blood as a result of cellular disorders (for the first time in 1954, by La Due, Wroblewski and Karmen for lactate-dehydrogenase, and Bruns for aldolase), the diagnostic use of enzymes widened to an extraordinary extent. The optic test described by O. Warburg in 1936, was a major contribution to these developments in that it provided the basis for a universally applicable method. At present, rapid progress is being made in enzyme activity measuring techniques. The investigation of isoenzymes and enzyme patterns has increased diagnostic specificity considerably.

Introduction of automation and electronic data processing

Since the early fifties, there has been a world-wide exponential increase in the number of analyses performed in clinical chemical laboratories. This has provided a stimulus to the development of mechanized analytical equipment. The first major outcome of this development was the introduction of the continuous flow analysis (L. T. Skeggs, 1957). The development which followed is still in full swing; for example, the introduction of the multichannel analyzer (1964) and the centrifugal analyzer (Anderson, 1969). In the strict sense of automation, however, these devices are non-automated machines because they have no feedback capabilities. The development of mechanized analytical equipment by itself would certainly not have initiated those fundamental changes which are presently becoming apparent through the concurrent introduction of electronic data processing (EDP) to clinical chemistry. Automation in conjunction with EDP opens up new dimensions for clinical chemistry. Not only has it led to improved quality of analyses and an increase in the frequency of analyses, but also, it has provided a means of examining large populations, of improving the presentation of data to the physicians, and finally, of introducing a large number of computer-assisted diagnostic aids.

NATIONAL AND INTERNATIONAL SOCIETIES AND PUBLICATIONS DEVOTED TO CLINICAL CHEMISTRY

Once clinical chemistry had achieved the status of a recognized discipline in its own right, national societies devoted entirely to the subject were formed in various countries. The first of these, the Societé Francaise de Biologie Clinique, was founded in 1942 in France and was soon followed by those in the Netherlands, the United States of America, the United Kingdom, and the Scandinavian countries. Societies of clinical chemistry have now been established all over the world. The most recent to be recognized by the IFCC are those of New Zealand, Egypt (1973), Chile (1974), Nigeria (1975) and Spain, Venezuela and Brazil (1976). At present there are 35 national societies †. In some countries, depending on historic and existing circumstances, the activities of the society may encompass other fields of clinical laboratory science, although it is recognized within the country as representing the national interest in clinical chemistry.

In 1952 the emerging national societies of clinical chemistry were organized as an international body under the auspices of IUPAC. The genesis of the International Federation of Clinical Chemistry was initially achieved simultaneously with the foundation of the Commission of Clinical Chemistry of the Section of Biochemistry of IUPAC. It was

† Argentina, Australia, Austria, Belgium, Brazil, Canada, Chile, Czechoslovakia, Democratic Republic of Germany, Denmark, Ecuador, Egypt, Federal Republic of Germany, Finland, France, Hungary, Iran, Ireland, Israel, Italy, Japan, Mexico, Netherlands, New Zealand, Norway, Poland, Portugal, Spain, Sweden, Switzerland, Union of Soviet Socialist Republics, United Kingdom, United States of America, Venezuela, Yugoslavia.

then anticipated that a time would come when national societies of clinical chemistry would form an International Federation of Clinical Chemistry affiliated with IUPAC as an Associate Member, but functioning as a fully independent entity. This occurred in 1964, with the founding of the IFCC by 18 national societies of clinical chemistry. The President of the Commission of Clinical Chemistry of IUPAC served simultaneously as President of the IFCC until 1967. At that time the independent role of the IFCC envisioned by its founders was initiated by the acceptance of an organizational framework which provided for separate activity of the IFCC while retaining its intimate and fruitful relation with IUPAC. Recognition by IUPAC of the marked growth and importance of clinical chemistry led to the elevation of the previous Commission of Clinical Chemistry to the status of a Section of the IUPAC in 1967.

The establishment of two apparently similar international organizations for clinical chemistry was important for a number of reasons. Members of the Section of Clinical Chemistry, IUPAC, though of different nationalities hold discussions and make decisions without regard of their national affiliations. In the IUPAC organization purely national matters are not considered as they may affect decisions on matters of international concern. It is on this basis that the IUPAC organization has achieved such remarkable results in bringing about international agreement on the fundamental subject of weights, measures, nomenclature and other matters familiar to all scientists.

The IFCC on the other hand, is a federation comprising at the present time the national societies in 35 countries. It is the IFCC which provides the meeting place for debate and reconciliation of national viewpoints on all matters relating to clinical chemistry. These include for example, problems of standardization, methodology, education, training, legislation, etc.

The two international bodies of clinical chemistry work very closely together and have many common members. For example, many of the national representatives of the IUPAC section also serve in the same capacity for IFCC. A number of topics are reviewed on a collaborative basis. One of their important functions is the sponsorship of international and regional congresses of clinical chemistry. The IUPAC organizations in clinical chemistry together with the IFCC have sponsored eight international congresses to date. Regional or national meetings sponsored by the IFCC are of more recent origin and include the First and Second Italian Congresses of Clinical Chemistry in 1971 and 1974, the Joint Meeting of the Austrian, German and Swiss Societies in 1971, the Czechoslovakian Congress of Clinical Biochemistry in 1971, and the First (1968), Second (1973) and Third (1976) Latin American Congresses in Mar del Plata, Argentina and Porte Alegre, Brazil and Caracas, Venezuela, as well as the Hungarian Congress of Clinical Chemistry in Pecs, Hungary, in 1974.

The United States of America gave birth to the first journal devoted solely to the speciality, *Clinical Chemistry* in 1952. This was soon followed by the international journal *Clinica Chimica Acta* by the *Annals of Clinical Biochemistry* (UK), *Zeitschrift fur Klinische Chemie und Klinische Biochemie* (Germany), *Bioquimica Clinica* (Argentina), and *Clinical Biochemistry* (Canada).

In some countries the polyvalent nature of the National Society is reflected in their publication, as in the case of the *Annales de Biologie Clinique* (France), *Diagnostyka Laboratoryjna* (Poland), and *Revista Facultad de Quimica Y Farmacia* (Ecuador).

CURRENT STATUS

Role of the clinical chemist in health assessment, diagnosis and treatment

The ever-increasing diagnostic use of clinical chemistry has resulted in a vast increase of service requests. Laboratories of any size show an increase each year of some 10-20% in the number of analyses performed. Certain of these analyses are of the greatest importance in the diagnosis of specific disorders. Other analyses find broad application in the investigation and treatment of diseases of the liver, kidney, gastrointestinal tract, the electrolyte changes occurring after trauma and surgical operations, the plasma protein disorders and most endocrine diseases. The increased application of enzyme and isoenzyme estimations in serum and other biological fluids have provided the clinician with powerful diagnostic tools for the assessment of diseases such as myocardial infarction, hepatitis and muscular dystrophy. The great increase in knowledge of the hereditary metabolic disorders has depended much upon the service supplied by the clinical chemistry laboratory. In phenylketonuria and other genetic disorders associated with mental deficiency this increased knowledge and the availability of laboratory control has enabled physicians to make more accurate diagnosis and render immediate appropriate therapy. In some situations this can avert the possibility of permanent mental disability. In the hereditary metabolic disorders, application of clinical chemical analyses to cultured cells or biopsy material is now frequently used in diagnosis. In this regard some of the diseases of muscle and the glycogen storage disorders should be mentioned.

Perhaps the outstanding success of the discipline in relation to treatment has been achieved in diabetic coma. Appropriate therapy monitored by clinical laboratory findings has sharply decreased mortality. The control of new drugs requires a great many clinical chemical analyses, first in animals and later in human applications. Cancer chemotherapy depends greatly upon such investigations. The determination of various serum enzymes and other substances has made possible the identification of carriers of hereditary disorders, the so-called heterozygotes. Parents can now be advised as to the likelihood of the metabolic disorders occurring in their children.

This rapid growth of clinical chemistry services both in range and in quantity has been accompanied by an increasing awareness of the need for the development of quality assessment and quality control procedures. The application of objective evaluation has revealed short-comings in the quality of clinical chemical results. Deficiencies in analytical systems and in individual performance have become evident. Thus, quality control procedures have provided a strong stimulus for the development of better analytical systems and training programs designed to produce clinical chemists competent and knowledgeable in both the technical and professional aspects of the discipline. As an additional consequence the quality of the work performed in clinical laboratories as well as the qualifications of the responsible individuals must meet minimal standards as defined by government and public health authorities in many countries.

It is clear that the supporting role of clinical chemistry in the delivery of health services has an interdisciplinary and continuously changing role. While it stands as an independent science today it has strong ties with physics, mathematics, biology, chemistry, biochemistry and medicine. As will be evident in subsequent discussion its practitioners must be well grounded in all of these subjects. Because of the social implications of health care delivery it is also essential that there be vigorous interaction between clinical chemistry, economics and sociology.

Role of the clinical chemist in medical education
Because of the obvious importance of the subject to medicine, the medical student, in most countries is introduced to clinical chemistry during his undergraduate years while acquiring both basic and clinical knowledge. Teaching of the subject usually begins as part of the presentation of biochemistry. While the clinical chemical component of bio-chemical courses varies from country to country, and indeed within a given country according to the interests of individual faculty members, it is usually the case that the fundamental aspects of biochemistry are illustrated and correlated with examples drawn from human problems. To an increasing degree these programs may include the presentation of patients. The role and extent of the student's laboratory involvement in clinical chemistry ranges from none in some instances to intensive courses of long duration in others. In all countries as the medical student advances to the completion of his education, exposure to the clinical aspects of biochemistry increases. Often this may be augmented by courses in clinical laboratory science followed in the later years by bedside discussions as well as correlated presentations of more formal lectures and clinical conferences. The essential goal is to provide the medical graduate with an integrated body of knowledge which will allow

the effective use of the laboratory for diagnosis and therapy. While detailed understanding of methodology is hardly required of the non-laboratory specialist, a feeling for the limitations inherent in measurements and results must be acquired. This can be provided in education in clinical chemistry throughout a career in medicine. It is therefore important that the clinical chemist who is in continuous contact with physicians, attain competence in teaching.

Clinical chemistry also has an important place in postgraduate medical education. The medical graduate who undertakes speciality training in one of the various divisions of medicine will require more in-depth knowledge in relevant areas of clinical chemistry. In obstetrics the course of normal pregnancy may require only basic knowledge of clinical chemistry but certain abnormal situations in which fetal integrity is in jeopardy may require extensive chemical testing of amniotic fluid samples. At such times clinical decisions are critically dependent on chemical information and the physician must have a thorough understanding of the biochemical and physiological principles underlying the wide range of chemical tests now available. Similarly in-depth knowledge of clinical chemistry is necessary for interpretation of the multitude of tests applied in the diagnosis and treatment of endocrine disturbances, gastrointestinal disorders and many other specialized problems.

In some universities, departments of clinical chemistry offer medical postgraduate general and specialized courses in clinical chemistry. They often provide resources and facilities for collaborative clinical-chemical research.

In addition to his contribution to undergraduate and postgraduate medical education, the clinical chemist also contributes to continuing education programs for medical graduates. It is his responsibility through formal lectures and the presentation of data at teaching and ward rounds in the hospital, to bring to the attention of his medical colleagues the status and pertinent advances in this field. Obviously for the education and training of medical graduates who wish to attain specialist status in the subject the clinical chemist and the clinical chemistry laboratory will be the central focus of the program.

The clinical chemist has an obligation to assist in the continuing education of laboratory technologists. This often involves the introduction and demonstration of new methods and new skills thereby providing a stimulus for continuing advancement and learning. In more formal academic settings the clinical chemist is expected to participate as a teacher in technology training programs.

Opportunities for the clinical chemist to participate in undergraduate teaching programs for chemists have been relatively infrequent. However, in countries where specialization in clinical chemistry is a primary responsibility of pharmacy education, the undergraduate teaching role of the clinical chemist may constitute a major part of his responsibilities.

Although most of the education and training programs in clinical chemistry have been established at the postgraduate level, new programs are evolving at the graduate or undergraduate level to an increasing extent.

Role of the clinical chemist in service

In order to fulfil his important function in the direct delivery of health services the clinical chemist, frequently acting in concert with his colleagues, is called upon to perform a variety of service functions. These include the supervision and direction of the clinical chemistry laboratory; the introduction of new procedures; the assurance of the quality and reliability of the analytical work which is performed; assistance in the interpretation of analytical data for the benefit of the patient; consultation with the medical staff with regard to the solution of special problems of a biochemical nature, and consultations regarding the social applications of clinical chemistry.

The supervision and direction of the clinical chemistry laboratory is a multifaceted activity. Regardless of future developments in automation and electronic data processing, the clinical laboratory will always be dependent upon the quality and performance of its personnel. Thus the procurement and utilization of an effective working staff will continue to be an important responsibility of the clinical chemist. Instrumentation, automation and electronic data processing have become critical components in the laboratory. The selection of the appropriate equipment must be based on an intimate knowledge of the characteristics of the workload, facilities, staff and economics relative to particular service requirements. The optimal analytical methodology for required tests is an indispensible component of laboratory performance, and its selection requires consideration of precision, accuracy and convenience, a complex process requiring a high degree of professional competence and experience. To achieve this and to provide a service which is of high quality, modern in scope and economical, the clinical chemist must maintain an active method development section prepared to investigate and test modern methods and equipment. This section fully investigates new methodology and automatic equipment prior to introduction into routine service. In some of the larger hospitals the methods development section may be large enough to require a full-time clinical chemist to take responsibility for its day to day operations. The clinical chemist in these situations often becomes involved in collaborative research projects with clinical investigators.

Even with a well-trained staff, well-maintained and properly selected instrumentation, and the selection of optimal methodology for a given environment, there is a continuing need for suitable programs of quality control to ensure satisfactory results on a continuous day-and-night, day-to-day service. The implementation of quality control programs has now become a legal requirement in many countries but in any event is requisite for the routine service operation of the clinical chemistry laboratory. The task is not completed, however, until the correct result is reported to the requesting physician. For this purpose the introduction of data management computer systems has become commonplace in the larger laboratories. The clinical chemist plays a major role in the selection and operation of this phase of the total laboratory performance.

When the clinical chemist is also a physician it is appropriate that the discharge of his service function may include interaction with his medical colleagues in the direct aspects of patient care. In this circumstance his specialized knowledge may be of assistance in furthering diagnosis and therapy. In many countries however, where non-medical clinical chemists serve in the capacity of director of the clinical chemistry laboratory this activity may be less direct. As called upon by physicians such an individual may offer his expert opinions for their benefit and guidance in relation to the laboratory results and their pathophysiological interpretation. In the same vein, and as fundamental understanding of human biochemistry is advanced, it becomes increasingly apparent that the clinical chemist coming from a science background can offer special insight to the clinician faced with the need to apply biochemical knowledge to unusual problems which arise in clinical medicine. These service opportunities often provide the stimulus for research investigations.

It is no longer possible to separate the day-to-day work of the clinical chemistry laboratory from its larger social context. The growing tendency to utilize 'biochemical profile testing' for preventive health as a part of health screening studies and surveys, carries with it an implicit judgement of the biological, economic, and technical significance of the tests which are performed. As part of his service role the clinical chemist is required to assist in reaching valid conclusions in these matters. Accordingly, his views must extend beyond his own laboratory setting in order to discharge properly his responsibilities in these broader aspects of service.

Role of the clinical chemist in research

At the postgraduate level the clinical chemist has been trained as a scientist. It may be expected, therefore, that the clinical chemist will maintain an active personal research program arising from his own interests, from his interaction with clinical colleagues and from the operation of the laboratory. A research environment in the laboratory provides good assurance of continued staff interest and optimal function. The stimulus for research activity is derived from continuous exposure to new forms of treatment with the need to understand their effects on humans, from the impact of the introduction of new analytical techniques and instruments on laboratory operation, and from daily interaction with clinical colleagues. Part of his professional function is to establish and maintain a laboratory capability which will serve as a focal point for research activities in a medical environment. The goals of early diagnosis, rational therapy, more exact prognosis and preventive medicine will be attained more readily as basic research is brought to the clinical setting. With current research has come the realization that an ultimate understanding of the disease process requires definitive information about pertinent reference value intervals for the concentration of biochemical constituents. These need to take into account ᴠariations associated with age, sex, diurnal variation, environment, diet, activity, ethnic origin and therapy.

Present problems in clinical chemistry
The rapid growth of clinical chemistry and the increasing scope of its involvement in new areas of clinical medicine has fostered many new problems. At the analytical level, the clinical chemist faces the need to provide more and more services at the highest attainable level of accuracy and precision. The continuing development and availability of better instrumentation and improved analytical systems has contributed in a major way to the solution of this problem. At the same time, however, improvement in analytical capability has created an even more pressing need for efficient ways of acquiring and handling laboratory data. The need to interface the computers effectively with clinical chemistry systems is complex and difficult. Undoubtedly the current technology will ultimately lead to more effective laboratory services and more meaningful laboratory data.

The physician has an increasingly difficult task of correlating and interpreting an avalanche of laboratory data with the clinical status of the patient. To assist this the clinical chemist must become more involved in the need to translate and codify the results of chemical analyses so that the information is more easily and more quickly assimilated by the clinician.

Additional problems of current importance to the practice of clinical chemistry include the necessity to attain agreement on nomenclature, quantities and units and to establish the effect of drug therapy on the values of laboratory tests.

These points underline the need to educate and train clinical chemists capable of functioning effectively. Present facilities are insufficient to meet the need for specialists which has developed in consequence of the expansion of medical services throughout the world.

The present deficiency of trained individuals also stems from the slow recognition of clinical chemistry as a unique multidisciplinary activity. It is only in the last two decades that a cohesive body of knowledge and activity has emerged to identify the speciality and permit a definition of its requirements. As a consequence, its present practitioners have come into the field from diverse backgrounds in medicine and physical science. Only with the present emergence of legal definitions of clinical chemistry practice has the stage been set by essential social acceptance for the initiation of definitive educational programs leading to speciality status. Likewise, the recognition of the important contribution of the laboratory to health care has favored the allocation of an increasing proportion of available resources to this activity.

Future trends
There is little doubt but that much of future progress lies in the continued development of automatic methods of analysis. Present advances in mechanization and electronics will surely yield to true automation characterized by built-in feedback resulting in appropriate adjustments for self-correction of the analytical device. The increased accuracy, speed and decreased cost of clinical laboratory testing will permit interval biochemical screening of whole populations to detect disease at its earliest beginnings and at a stage when it may be reversible.

Such an advance will also require medical knowledge of biochemical and other indices precedent to disease. The advent of computers in the medical environment presages the time when the patient history and record will so direct the process of automated analysis that maximal significant information for diagnosis and treatment will be obtained in the first biochemical examination of the specimen.

A trend toward reorganization of clinical laboratories is now evident. The development of regionally centralized laboratories has been encouraged by the increased ease of transportation and sample handling, the increased cost of capital equipment and the higher labor costs associated with low-volume testing. We are seeing a new generation of inexpensive instruments and methods, capable of performing emergency tests with great speed at the bedside or in the hospital laboratory, and simultaneously the establishment of centralized facilities for handling high-volume, less urgent tests efficiently and economically. The parallel development of data processing technology furthers the concept of laboratory service centralization by the consequent enhanced transmission and reporting of test results.

Developments in medicine and surgery interlock with the advances to be expected in the clinical chemistry laboratory. The increasing use of tissue and organ transplantation in medicine has brought with it a major requirement for the development of immunological procedures in the biochemistry laboratory. The presently available basic knowledge about the structure and composition of the immunoglobulins, together with a better understanding of their functions in health and disease, will undoubtedly lead to routine clinical chemical testing for these substances in the near future.

During the past few years the development and application of a variety of immunological techniques to problems in clinical medicine has taken a position of major importance. Radioimmunoassay in particular, is today one of the most rapidly-growing areas in laboratory medicine. While its clinical application has been primarily to the determination of various protein and peptide hormones, radioimmunoassay methods are being extended very rapidly to the analysis of many other biological substances. Thus, procedures have now been developed for assay of therapeutic agents, notably the cardiac glycosides, steroids, enzymes, drugs of abuse, vitamins and other metabolites. The outstanding value of radioimmunoassay methods is their great sensitivity, specificity and simplicity. Undoubtedly in the future they will displace many conventional chemical procedures. Their availability in the clinical laboratory will facilitate diagnosis, the evaluation of therapy and the definition of changes precedent to disease.

Possibly the most challenging future potential is the development of genetic engineering to correct inherited disease. Directing this most fundamental biological power into avenues of individual and social benefit will test to the utmost the capability of our educational systems to instill wisdom along with learning.

3

The Education and Training of the Clinical Chemist

ROLE OF THE CLINICAL CHEMIST

The clinical biochemist studies and advises on chemical and biochemical processes in the organs, tissues and fluids of the human body and the effect of disease and drugs on these processes. He organizes the routine service of analytical and functional tests; devises and tests new methods of investigating disease processes; and applies to clinical problems all modern chemical technique*. The functions of the clinical chemist have also been analyzed in terms of 'basic research, applied research development and laboratory service'† The same activities are embodied in both descriptions.

Although the distribution of an individual's efforts in the position of clinical chemist will vary depending on circumstances, any satisfactory program of education and training must encompass these possible functions. No single program of formal study can be specified which will automatically insure the attainment of these goals. Requirements of law and custom, the leadership of gifted individuals, the presence of suitable physical facilities and a climate of interdisciplinary intellectual activity will shape the pattern which evolves in any given place. The fundamental requisite for each function of the clinical chemist, however, can be identified.

ASPECTS OF EDUCATION

Basic research

The key to productive study of fundamental phenomena is to be found in an attitude of mind. The outlook which questions, challenges, probes, tests and verifies cannot be taught, but must be learned. Over the years general agreement has been reached that formal doctoral study is the most effective method of providing this learning opportunity. Its requirement for the individual delineation of a problem and the study, critical interpretation and defense of conclusions based upon research sets this pattern. For those who come into the field by other routes, an alternative to formal education in science can be provided by an opportunity to participate intensively in research as a means of attaining an essential mental and work discipline. The single certainty for future

---

* *Hospital Scientific and Technical Services Report of the Committee 1967–1968.* Chairman: Sir Solly Zuckerman, Her Majesty's Stationery Office, London, 1968, p. 51
† Kinney, T. D., and Melville, R. S. (1969) The Clinical Laboratory Scientist, *Lab. Invest. 2, 382.*

basic advances in clinical chemistry is that they will stem from developments in many fields. The tools to be provided in the educational program, therefore, must be broad and heavily weighted in the direction of the physical and biological sciences. Throughout the program a deliberate effort should be made to extend the theory and practice of the defined systems of physics and chemistry to the unique multivariate problems of living things. A subsequent orientation to medicine and human disease can be readily built into such a foundation.

The detailed nature of a suitable curriculum for clinical chemistry will vary with the immediate circumstances of the social and institutional environment. Flexibility in the educational form is also to be encouraged in order to accommodate to the rapidly changing character of the subject. Basic essentials, however, surely include mathematics and statistics, general, analytical, organic and physical chemistry, a thorough grounding in biochemistry and physics including optics, electricity, electronics, fluid dynamics and thermodynamics. Fundamental biology, cell biology and morphology, physiology, microbiology, pharmacology and pathology are important. Some experience at the bedside in order to develop a feeling for the manifestations of disease, its natural course and the response to therapy could be most valuable. Few present programs provide the above outlined breadth of education in a single established curriculum. As most clinical chemists will come into the field with a background in the sciences, pharmacy, or medicine an opportunity should be provided in each case for completion of requisite studies.

Applied research and development

One essential function of the clinical chemist is to provide the link whereby technological improvements are harnessed to facilitate the work of the clinical chemistry laboratory. Too often in the past the clinical chemist has had to utilize as best he could, equipment designated for other tasks. Advancement in the technology of the clinical chemistry laboratory will come most rapidly when the clinical chemist initiates the changes dictated by his specific requirements. It may no longer be possible even for a gifted individual to master all the technical elements involved in the design and fabrication of modern equipment. The clinical chemist, however, by virtue of his broadly based education and intimate knowledge of the problems to be solved is the natural leader of activities directed toward continued technical improvement and modernization of the laboratory. The development of new equipment, procedures, and concepts implies a concomitant unbiased critical evaluation under the operating work conditions of the clinical laboratory. The scientific attitude induced during basic research will assure the recognition by the clinical chemist that change is not necessarily progress and that new methods and new ideas require careful appraisal.

18

Service

A major function of the clinical chemist is to provide the best possible supporting service to other professional colleagues. Interest in basic research, in applied research and development, and a judicious application of management techniques will all contribute to this objective. Most significant, however, must be an acute awareness and definite involvement in the problems of the patient, the clinician and the social implications of his work. The non-medical scientist will learn this essential attitude by intensive participation in the clinical activities of a medical center including attendance at the various medical conferences, the teaching clinics and observation of the bedside care of patients. He will thus learn of the needs of patients, the language of medicine and the problems faced by his clinical colleagues.

Management

Neither a strong basic research interest, an active development program, a service-motivated outlook, or all these attributes together, will insure a smoothly functioning, efficient organization. To attain this goal, the clinical chemist must also be a good manager. He needs to weld technical and medical developments into realistic objectives appropriate to his organization. Inevitably he will face and must solve the myriad problems associated with the interactions of individuals in a group. Finally it will be an essential part of his task firmly to establish the role of the clinical chemistry laboratory in the general scheme of medical services so that its potential contributions are realized and its activities adequately supported. The methods of management can be provided by formal teaching but there is no satisfactory substitute for training in the actual laboratory environment. Courses in management are not often available in the university scientific and medical curriculum. They can be provided through the co-operation of the schools of economics of most large universities and are frequently available in organized symposia. The education of the clinical chemist should include an opportunity to benefit from such programs.

Participation in the management and direction of a well-organized department of clinical chemistry is an essential part of the training of the potential director. Problems of personnel, allocation of budget, evaluation of future needs and the development of the resources required for technical advance are all aspects of laboratory management which can best be mastered by first-hand experience. This education for direction of the clinical chemistry laboratory needs to be provided. Wherever possible, the graduate student should also gain some lecturing experience, a very important tool in communicating ideas to others.

SUPPLEMENTARY EDUCATION AFTER GRADUATION FROM PROGRAMS IN MEDICINE, SCIENCE AND PHARMACY

Because of the interdisciplinary nature of clinical chemistry, the background provided by the usual curricula in medicine, science and pharmacy, may require specific supplementation. Suggested programs appropriate to the education of candidates entering the field with these differing backgrounds are presented in the following sections.

It would be expected that the examination of all graduates in clinical chemistry, whether they start from a background of medicine, science, or pharmacy, will be subjected to the same or equivalent examinations with perhaps recognition of the nuances conferred by their previous specialization.

For the medical graduate

The medical graduate by virtue of his clinical training and experience has assets which enable him to exert a direct influence on his medical colleagues when they place excessive and perhaps unreasonable demands on the laboratory. He is usually more concerned with clinical interpretation of laboratory results than other members of the team and may devote more of his time and activities to communication with other medical officers. He is also more likely to have administrative responsibilities outside the clinical chemistry laboratory. This does not diminish, however, the need for basic training and education in scientific disciplines related to clinical chemistry.

Basic sciences

In addition to basic knowledge in general, organic, physical and radiochemistry and genetics, the medical candidate may require additional theoretical and practical training in analytical chemistry, statistics and biochemistry including human biochemistry. The need for training in physics (with the exception of electronics), biophysics and higher mathematics is perhaps less important, and the essentials may be acquired in the course of the aforementioned studies.

Clinical chemistry

The training in clinical chemistry should comprise specific theoretical and practical training in the biochemistry of clinical medicine. This involves the interpretation of both clinical and laboratory data. It requires essential knowledge of analytical clinical chemistry, including the setting up and evaluation of methods, analytical techniques, quality control, instrumentation and mechanization. Much of this training may be organized as lectures, tutorials and practical work in a university department of clinical chemistry. Understanding of metabolic disorders and clinical interpretation of laboratory data should be acquired in a clinical setting.

Training in research can only be acquired through actual participation

20

in research activities and eventually the trainee should conduct a research project of his own.

Laboratory management

The planning of work, laboratory design, budgeting, work simplification, data handling and other aspects of management are learned best by the experience gained in a clinical chemistry laboratory under the guidance of an experienced clinical chemist. This may require supplementation by available formal educational programs.

Subjects related to clinical chemistry

Both theoretical and practical training should be given in hematology, clinical physiology, endocrinology and the use of isotopes. Practical experience should be gained preferably in a specialized laboratory or department. Basic training in serology, immunology, microscopic pathology and pharmacology is part of the medical course in most universities. No extra training seems to be required, except possibly in immunochemistry and toxicology.

Clinical medicine

It is important that the trainee will have spent a total of at least six months in clinical wards, preferably internal medicine or pediatrics, acquiring experience in the application of laboratory findings to patient care.

Length of training and final examinations

Two to three years of the training should be carried out in a university department of clinical chemistry or in an appropriate clinical chemistry laboratory. A diploma or certificate should be issued after the candidate has passed an examination conducted by a competent and representative board.

For the science graduate

The scientist with a higher qualification approaches the subject of clinical chemistry with the advantage of a strong background in research. Recognizing that this may have been highly selective and specialized it would be important to ensure that each candidate had acquired adequate fundamental knowledge and skills in the basic sciences specified in a previous section. Special study of the principles of electronics and training in the use of computers are likely to prove helpful.

Medical knowledge

It is essential that the science graduate attains: knowledge of fundamental biology, cell biology, cell morphology, physical biochemistry, expert knowledge of human biochemistry and appropriate knowledge of other aspects of human biology; an appropriate knowledge of medical nomenclature; a manifest understanding of the etiology and pathogenesis of diseases showing chemical disturbances; familiarity with drug therapy, drug metabolism and with chemical and hematological effects of drugs; knowledge from practical experience of the function of hospital wards and outpatient departments; familiarity with laws and rules regulating public health services and hospitals, and medical practice and its ethics.

The method by which the student obtains an understanding of these topics, whether by formal courses, precept or self education, will vary with local circumstances and opportunity.

Clinical chemistry, laboratory management and subjects related to clinical chemistry
These require the amplification described previously in the case of the medical graduate.

Length of training and final examinations
Two to three years of postdoctoral education depending on the previous preparation of the student, should be sufficient to impart the required background. Clearly the appropriate environment is a clinical chemistry laboratory in an active university center of the highest caliber. As in the case of the medical graduate the satisfactory completion of study should be attested by passing an examination before a competent and representative board.

For the pharmacy graduate
To a greater extent than in medicine and the sciences, education in pharmacy is of variable quality in the countries of the world. Educational programs for pharmacists seeking specialist recognition in clinical chemistry will thus vary from country to country according to the content of undergraduate and postgraduate curricula in pharmacy. Pharmacists who receive five years of undergraduate training are generally well trained in the basic sciences. Their curriculum should also include in-depth education and training in human biology, bioanalytical techniques applied to human pathology and general clinical chemistry. Others will require supplementation in basic sciences as outlined for postgraduate education in clinical chemistry. Postgraduate education in pharmacy, highly orientated toward clinical chemistry, is available to pharmacy graduates qualified as described previously. Such candidates should acquire training at advanced levels in analytical chemistry, clinical chemistry including instrumental analysis, pharmacology, toxicology, and basic training in medical and surgical pathology, immunochemistry and hematology.

In all of these programs varying degrees of supplementation will be required for pharmacy graduates proceeding to specialist status in clinical chemistry as described for graduates in medicine or sciences.

As in the case of medical and science graduates, a program of education and training leading from a pharmacy background cannot be considered to be completed until the qualified candidate has passed an appropriate examination before a competent and representative board.

Continuing education experience in the practice of clinical chemistry

In addition to the completion of an approved educational program, the candidate in clinical chemistry needs practical experience in a clinical chemistry laboratory. A period of two to three years as an assistant in a laboratory headed by a qualified director can provide this background, and should be required as a condition of the recognition of specialization.

The primary goal of education particularly at the university level, is to develop in the student thorough systematic training and stimulation of moral and intellectual faculties and a sense of intellectual self sufficiency. Education in these terms has a life-long effect and ensures continuing scientific progress. It is essential, nonetheless, that the completion of formal qualifications is not the end of education.

The question as to whether a director should be subject to re-examination to establish continued qualification at regular intervals is an unsettled and difficult one. It has not been instituted in any country as yet, and is not customary in other professions.

EDUCATION OF TECHNICAL STAFF

Advances in instrumentation, electronics, on-line computerization, chromatography and many other methods of special chemistry have resulted in complex chemistry and technology. If full advantage is to be taken of these advances, a greater number of more specialized highly skilled and more knowledgeable senior laboratory workers will be required.

Senior chemistry technologist

The supervisor of the clinical chemistry laboratory, working under the director, applies special skills and knowledge of chemistry to day-to-day routine service operations. Individuals with university level education in chemistry may prepare themselves for this role by undertaking some formal graduate school education with special attention directed toward training in the clinical interpretation of laboratory data, electronics and instrumentation, biochemistry, biostatistics and quality control. A further extension of education and training may be desirable in such subjects as clinical pathology, laboratory practice in clinical chemistry, computer science, calculus, endocrinology or hematology. Individuals with this kind of background would be suitable for supervisory roles, or for taking charge of special and routine chemistry laboratories, depending on the size and function of these laboratories. Alternative routes are available by which the technologist may attain supervisory status. The combination of continuing education programs, within or external to the laboratory, with specialized laboratory and management experience provides a mechanism for this advancement.

Senior non-chemist technologist

It has become obvious that the growing complexity of the clinical chemistry laboratory will demand the services of individuals competent to supervize specialized functions. Computer technology applied to problems of communication, process control and data handling, will require trained staff. The maintenance, standardization, design and development of analytical instrumentation now requires skills possessed by instrumentation specialists. While the degree of diversification of function and responsibility within the laboratory will depend on individual circumstances, the trend towards laboratory centralization will bring with it an increasing involvement of non-chemical specialists. Even though chemical science continues to remain the basis of the work of the clinical chemistry laboratory, it is hoped that rewarding positions of technical responsibility will be established for non-chemical professionals. Their education and training in addition to certain fundamentals of chemical science can be expected to include a greater depth of training in instrumentation, statistics, basic principles of electronics, computer mathematics, research techniques, data processing and systems management. They will be expected to accept greater responsibility for certain and complex tasks and for the supervision of other laboratory personnel in their performance.

Technologist

The delivery of health care is necessarily achieved by a variety of means, ranging from private practice, to large and small hospitals, polyclinics and very limited rural facilities. In an analogous fashion the needs for laboratory services and the requirement for personnel for these different facilities is also variable. Two classes of laboratory technologists can be recognized; those who function entirely in clinical chemistry and others whose duties also include hematology, blood banking, microbiology, and parasitology. In general the individuals working only in clinical chemistry are to be found in the larger institution while the 'polyvalent' technologist functions in a smaller laboratory installation.

To some degree the training of these two groups is different.

The chemistry technologist

In recent years educational programs have been advocated and instituted for technologists who will function entirely in clinical chemistry. These have been based primarily upon a college or university degree program with major concentration in chemistry. All the usual requirements are met for chemical education including general, analytical, organic and physical chemistry. In addition these programs include considerable effort devoted to instrumentation, computers, physics, and a basic background in human biology. Additional training is provided in a hospital clinical chemistry laboratory in order that the student may learn to apply theoretical knowledge in a practical setting. The programs are generally of three to four years duration following completion of secondary school education. Individuals who have completed such programs are recognized in various countries as technologists for the speciality of clinical chemistry.

4

Conclusions and Recommendations

CONCLUSIONS

(1)   Clinical chemistry is an interdisciplinary subject with strong roots especially in the fields of chemistry and medicine. It has emerged as a definitive body of knowledge and practice.

(2)   The subject is rapidly expanding and is providing essential support for the delivery of health care as well as extension of basic knowledge in the health sciences.

(3)   These facts have been increasingly recognized by the promulgation of laws and regulations governing the conduct of clinical laboratory analysis and the qualifications of the individuals engaged in this work.

(4)   Clearly an insufficient number of trained specialists is presently available to meet the current and anticipated needs in this speciality.

(5)   While proficiency in the subject may be achieved from an initial background in medicine, the physical sciences and pharmacy it is now recognized that a definitive program of education, training and experience is requisite to attain the specialist status which can be codified by examination, certification and licensure.

RECOMMENDATIONS

(1)   That the education, training and experience of the director and senior professional staff of the clinical chemistry laboratory meet certain minimal standards. Education should provide an advanced knowledge of chemistry and biochemistry; basic knowledge of mathematics, statistics, physics, human biology and genetics; a grasp of physiology, histology, pathology, pharmacology, toxicology and medicine with extensive understanding of chemical pathology; and a working knowledge of modern analytical instrumentation and electronics, possibly extending to computer science. Training in research methodology needs to be balanced by a background in personnel and administrative management. Extensive experience in a suitable clinical chemistry laboratory is an essential part of the educational process.

(2)   That universities be encouraged to provide the educational facilities and environment for professional development in clinical chemistry and that this be fostered by the creation of chairs devoted to the speciality.

(3)   That institutions responsible for postgraduate activities in the

professions establish programs and courses for continuing education and training in clinical chemistry.

(4) That appropriate international and governmental organizations promote and support programs in this field.

(5) That in the context of this report, governments and their regulatory agencies establish rules for the practice of clinical chemistry and the certification of specialists in the field.

(6) That international organizations concerned with health care such as WHO, UNESCO, IFCC, IUPAC and others, recognize the world-wide need for education and training of personnel for clinical chemistry laboratories and that they foster, promote, encourage and support programs to accomplish these purposes.

27

5

Worldwide Clinical Chemistry

The following section provides an individualized account of the status of clinical chemistry in many countries. It should be recognized that the information may be incomplete and already partly out-of-date since it was collected for the most part during the period 1972-1976. As for any subject in a state of very rapid change the detailed existing situation in a given country may vary from the material presented. Nonetheless the total picture which has emerged from these accounts reasonably reflects the international status of clinical chemistry.

ARGENTINA

History
In the Universidad Nacional de Buenos Aires in 1919 a curriculum leading to the 'doctorate in pharmacy and biochemistry was created for the study of the components of the blood, urine and other humours as well as the action of medicaments in the body'. As a first attempt this career was planned for graduates in pharmacy. It had a duration of two years with a program composed of ten subjects, and it finished with a research study and thesis. This program required seven or eight years study after secondary school. This career opportunity was subsequently provided in the universities of the country: La Plata, Cordoba, Tucuman and Litoral. The creation of this career, considering the fields it embraces was achieved for the first time in the world in an Argentine university. We are indebted to the interested and tenacious efforts of Professor Juan A. Sanchez for its development. In 1959 the Universidad Nacional de La Plata established the degree of 'licentiate in clinical biochemistry'. Afterwards six other universities organized programs for degrees in biochemistry separately from those in pharmacy.

Today all the universities of the country grant a degree in biochemistry or clinical biochemistry. The Universidad Nacional de Buenos Aires is a special case, for it has two faculties, that of 'pharmacy and biochemistry' and that of 'exact, physical and natural sciences'. Graduates of the latter program are considered as professionals qualified to work in clinical analysis. The course in pharmacy and biochemistry comprises a period of six years. Upon completion of a

28

research thesis the degree of doctor in biochemistry is awarded. In the faculty of exact, physical and natural sciences, after six years of study, the degree of doctor in chemistry and biological speciality is attained. In Argentina, since the creation of this career in 1919, the performance of clinical analysis has been limited to the field of chemistry and sporadically within the medical field. The most important problem has been the multiplicity of the kinds of degrees awarded by the Argentine universities to provide qualification for carrying out clinical analysis. An additional problem has been the granting of degrees for different functions. For this reason the Ministerie de Salud Publica de la Nacion, Buenos Aires, established professional qualifications for the performance of clinical analysis. The university, after having consulted with the faculties of medicine, exact, physical and natural sciences and pharmacy and biochemistry, established on May 8th, 1967 that for the career in clinical analysis the following subjects would be required:

> Mathematics (including biostatistics)
> General chemistry
> Physical chemistry
> Inorganic chemistry
> Analytical chemistry
> Organic chemistry
> Biological chemistry
> Microbiology, parasitology and immunology
> Concepts of anatomy (macro and microscopic)
> Physiology
> Clinical analysis (including physical, chemical,
>   biological, bacteriological and parasitological)
> Practice: annual course of intensive practice in a
>   clinical laboratory, laws of professional practice
> Toxicological analysis

According to this criterion of the Universidad Nacional de Buenos Aires, only the biochemists or biological chemists fulfil the requirements for the practice of biochemistry. A medical doctor is not considered as qualified. The curriculum of the Universidad Nacional de La Plata for the 'licentiate in clinical biochemistry' is as follows:

> Calculus 1
> Calculus 2
> Algebra, analytical geometry
> Physics 1
> Physics 2
> Physical chemistry 1
> Physical chemistry 2
> General chemistry

29

Inorganic chemistry
Analytical chemistry 1
Analytical chemistry 2
Organic chemistry 1
Organic chemistry 2
General biology
Botany
Microbiology 1
Microbiology 2
Anatomy
Physiology
Histology
Bromatology (nutrition)
Biological chemistry 1
Biological chemistry 2
Biophysics
Clinical analysis 1
Clinical analysis 2
Toxicology and legal chemistry
Mycology and parasitology
Pathologic anatomy and physiology
Pathologic biochemistry
Clinical biochemistry

The following Table provides a list of the universities, faculties or colleges and degrees which are equivalent to those of the Universidad Nacionál de Buenos Aires.

University	Faculty or College	Degree
Buenos Aires	Facultad de Ciencias Exactas, Fisicas y Naturales	Doctor or Licenciade en Bioquimica
Buenos Aires	Facultad de Farmacia y Bioquímica	Doctor or Licenciade en Bioquimica
La Plata	Facultad de Ciencias Exactas	Licenciade en Bio- quimica Clinica
Litoral	Facultad de Ingenieria Quimica	Bioquimica
Rosarie	Escuela de Bioquímica y Farmacia	Bioquimica

Cuye	Facultad de Ciencias	Licenciade en Bio- química
Sur	Departmento de Quimica e Ingenieria Quimica	Licenciade en Bio- química
Tucuman	Facultad de Bio- Química Quimica y Farmacia	Bio- química
Cordoba	Institute de Ciencias Quimicas	Bio- química Clinica

Legal status

In Argentina each province has its own legislation for the practice of clinical biochemistry. A national law which operates in the federal district and national territories specifies that the clinical analysis may be performed by medical doctors, biochemists, and chemists, who have been qualified. In the province of Buenos Aires, which has a population that is one-fourth that of the whole country, the legislation allows the practice of clinical analysis to those professionals who have completed the requirements of the Universidad Nacional de Buenos Aires as specified in 1967. In addition the practice of biochemistry is specified as being distinct from any other speciality of the so-called 'healing art' including pharmacy. The practice of sharing professional honorariums is also forbidden. The regulation of professional practice is entrusted to two types of associations by the state. One, usually named college or council controls the matriculation, manages aspects of ethics and morals and defends the interests and health of the population. These institutions are created by law. Civil associations which specifically defend the interests of the professionals in respect to their economy and working norms, also organize meetings and conventions for the scientific improvement of their members. These associations may be local or regional and may group within a province to form a Federation. The Federation may associate to form a National Confederation. In Argentina there are two Confederations, one is the Argentine Biochemist Confederation and the other the Clinical Biochemical Confederation of the Argentine Republic. They are geographically distinct but have more or less similar purposes.

Postgraduate teaching

The universities and the professional associations are constantly organizing courses and scientific meetings for postgraduate education. In addition there is a College Hospital in Buenos Aires where practical teaching is given to students and postgraduates. In general the scientific level of the biochemists is high. There are biochemists who specialize in all of the branches of modern biochemistry: enzymes, hormones, proteins, radioimmunoassays, hematology, bacteriology, etc. Interchange of information in broadly based Latin American courses is under consideration. The use of Spanish as the language of instruction and locally available equipment is essential to ensure optimal results and high attendance by participants. Education in a foreign tongue or with too highly specialized equipment is ineffective. This point is given special emphasis to ensure that national societies, Latin American confederations and the IFCC organize plans that could fit local circumstances.

Professional organizations
As a result of the First Latin American Congress of Clinical Biochemistry that took place in Mar del Plata in December, 1968, the Clinical Biochemical Confederation of the Argentine Republic was established. The Confederation tries to promote the collective interests and unity of clinical chemists. It strives to ensure that competent professionals with university qualification perform analytical testing. It attempts to achieve uniformity in university degree programs. It promotes professional ownership of laboratories and sets tariffs for private activities, social work, clinics and sanatoriums. In this manner the professional members of the organization are freely engaged in activities, social work, that of mutual societies, clinics and sanatoriums. Members of the social and mutual societies pay a share in relation to their incomes and a similar part is paid by the employer. Those affiliated have the right to complete medical, radiological and laboratory services, and usually to a percentage of the cost of medicines. The Confederation of Biochemists of the province of Buenos Aires performs these functions for this region of the country.

Practice of clinical analysis
The practice of clinical analysis in Argentina is performed in the official hospitals, private clinics, sanatoriums and private laboratories. The Argentinian hospitals are free and reserved for the people with poor incomes. Salaried professionals are in charge of these facilities. Some of the clinics and sanatoriums have their own laboratories for clinical testing of ambulatory and bed patients. Most analyses are carried out in private laboratories. Argentina does not have the equivalent of a technologist degree, education is at the professional level. It is hoped that this viewpoint can be generally acceptable in Latin America. That this may be possible is indicated by the action of the national members of the Latin American Confederation of Clinical Biochemistry on the occasion of its first ordinary assembly. At that time, having considered the circumstances inherent in the practice of their profession, they concluded that:

(1)   Clinical biochemistry is a very important branch of human knowledge that effectively supports medicine in diagnostic prediction and the treatment of disease.

(2)   By virtue of their knowledge, the professionals in this activity constitute irreplaceable support in the preservation of health.

(3)   The rapid evolution of learning in this field makes it impossible to practise this profession without university and continuous postgraduate education.

For these reasons the Latin American Confederation of Clinical Biochemistry proposed that:

(1)   Only university professionals who have received scientific instruction in clinical biochemistry be allowed to perform analyses.

(2)   It constitutes the maximum benefit for the health of the Latin American people that the government of their respective countries adopt appropriate rules to govern this activity.

AUSTRALIA

History

Clinical chemistry in Australia can be traced in its earliest form to the rudimentary laboratory at the Coast Hospital in the years prior to the First World War. By the 1930s hospital clinical chemistry laboratories and individuals specializing in the subject were to be found in a number of locations including the Royal Perth Hospital, the Sydney Hospital, the Royal North Shore Hospital in Sydney and the Royal Adelaide Hospital. The names of Dr. H. S. H. Wardlaw and Professor R. Lemberg are associated with the development of the speciality in those years. Analysis of urine and of blood for sugar, residual nitrogen and urea were the starting points for the development of the subject.

In the 1950s significant changes occurred with the appointment of PhD clinical chemists to hospital laboratories. The need for professional interchange and communication led to the formation of the Australian Association of Clinical Biochemists (AACB) in 1961 with an original membership of 70. From its inception the AACB undertook efforts to improve the quality of work performance. Surveys for quality control were carried out by distribution of samples in conjunction with the College of Pathologists of Australia. By 1969 the membership of the AACB, now totalling about 270, had been divided by a Board of Examiners into a highly-trained group of 126 'Foundation Fellows' and 'Foundation Members'. Subsequently fellowship and membership examinations have been conducted annually. The first chair for clinical chemistry was created in 1968 by the University of Western Australia.

Legal status

National regulations
There are no national regulations relating to clinical chemistry. This is left to individual states.

Local regulations
In Australia the position on local regulations varies greatly from state to state; anybody can set up a laboratory in the State of Victoria, take blood from patients and carry out analyses which will be accepted by general practitioners. On the other end of the scale is Queensland where according to state law, only a medically-qualified registered practitioner can even call himself a biochemist. Blood can be taken by non-medical personnel in the hospitals, but the medical superintendent or the medically-qualified pathologist takes the final responsibility.

There are virtually no legal requirements for non-medical personnel working in hospital laboratories in Australia. It is up to the hospital administration to decide which type of person they wish to employ and in what capacity. This is usually regulated at least in the government hospitals, by industrial awards according to the schedules laid down by the respective states.

Accreditation of laboratories
A scheme for the accreditation of laboratories in Australia is under discussion between the Australian government, the states and the professional organizations.

Education

University programs
Most clinical chemists in Australia have graduated in chemistry or biochemistry and obtained their specialist training during employment in hospital biochemistry laboratories. In one or two universities, notably Western Australia, it has been possible for hospital staff members to engage in part-time postgraduate research work with a clinical chemistry bias, leading to a PhD degree. A Master of Science by course work will be started in the University of Western Australia in 1976.

Professional bodies

Royal College of Pathologists of Australia
All medical graduates receive some training in clinical chemistry during their medical course. This is given however in most universities through the departments of medicine or pathology.

If a medical graduate decides to become a pathologist he can join the training scheme of the Royal College of Pathologists of Australia. Under this scheme he is employed as 'trainee pathologist' in one of the large recognized hospital laboratories, spending about one year in histopathology, hematology, biochemistry and microbiology, studying a prescribed course of reading and sitting for an examination in each discipline. He is then regarded as a general pathologist. A few of these specialize in chemistry and sit for a higher level of chemical pathology.

Australian Association of Clinical Biochemists

The association has introduced a grading of its membership and appointed a board of examiners. A degree and an active interest in clinical chemistry allows entry as an associate. After three years of experience, an examination is available for the award of Membership (MAACB) and after ten years a higher examination for Fellowship (FAACB). Schemes of study for Membership and Fellowship candidates have been drawn up by the Committee on Education. In some centers teaching sessions are held to help prepare graduates for the examinations.

Place of clinical chemistry in hospital service

The administration of hospitals in Australia is complicated by the political system of the federation of states and under the recently introduced Medibank agreements hospitals are funded jointly by Commonwealth and state. The Department of Repatriation, centrally controlled, concerned with the health and welfare of members of the forces, especially after discharge, also runs a hospital system with its own laboratories.

Types of laboratory

(1) *Large hospital laboratories.* These usually are part of the teaching hospitals attached to universities, but not controlled by them. They are controlled either by the state government, a religious community, or are independent.

The laboratories are mainly under the administrative control of the hospital board, and are usually directed by a medically-qualified pathologist. In some hospitals, e.g. the Royal Perth Hospital (Western Australia) and the Prince Henry Hospital in Sydney, each laboratory discipline is separately directed and the head of each is directly responsible to the hospital administration. The head of clinical chemistry may or may not be medically qualified.

Many of the heads of biochemistry divisions or departments have university appointments to teach medical students.

The Repatriation Hospitals in Perth, Adelaide, Melbourne, Sydney, Brisbane and Hobart have large laboratories under the direction of a pathologist to whom the head of the biochemistry division is responsible. All these biochemists are science graduates, but only two of them have PhD degrees. One of these PhDs recommends methods and instruments for all biochemistry in the Repatriation Department.

(2) *Small hospital laboratories.* These are usually under the direction of a pathologist, frequently without specialist biochemical training, under the administrative control of the hospital. The pathologist controls the biochemical work, but it is carried out usually by a technologist responsible to the pathologist.

Place of clinical chemistry in non-hospital services

Private pathologists

Private pathologists are in all Australian states. The medically-qualified pathologists carry out a wide range of diagnostic laboratory work. Some have very large establishments employing scientifically-qualified biochemists, but the majority rely on technologists and technicians to carry out their chemical work.

Commonwealth health laboratories

These are set up by the Commonwealth Health Department in various large country towns to provide diagnostic services to the medical practitioners in the area who send the specimens in for analysis. They are usually directed by a technologist and carry out all types of diagnostic laboratory tests. They are under the control of a central office in the Federal Capital, Canberra. There is a scientifically-qualified biochemist in this office who has nominal control over the chemical work carried out in the various laboratories.

General practice laboratories

There is an increasing number of small laboratories, run by young technologists, or science graduates, attached to private general practice groups.

Screening laboratories

Recently at least three groups have started variants of multiple health examination laboratories. Multiple channel analyzers and computers are being installed, taking advantage of the financial structure of the Commonwealth benefit scheme under which considerable profits are possible when large numbers of patients are available. This is causing concern among the more conventional pathologists and biochemists.

Place of clinical chemistry in medical education

As described previously, all medical graduates receive some training in clinical chemistry during their medical course. The teaching of clinical chemistry to undergraduates in medicine is not directed towards producing professional clinical chemists but is intended as part of general medical teaching.

Place of clinical chemistry in non-medical education

Academic teaching of clinical chemistry in science, either at undergraduate or postgraduate level is virtually non-existent in Australian universities, yet graduates in science constitute the majority of professional clinical chemists.

In Australian universities, diversion of scientists from clinical chemistry commences during undergraduate days when students are discouraged by the staff, through ignorance of the opportunities, from working towards a career in clinical chemistry. The division becomes practically irrevocable in those students who proceed to senior degrees in biochemistry or chemistry, for informed employers are unlikely to employ at the relatively high salaries prescribed by award, these MSc and PhD graduates who have no experience and little useful knowledge of clinical chemistry.

It is common, if not usual, for a student to graduate BSc without having even heard of the large numbers of graduates employed in hospitals. This implies shortcomings in the work of student advisers, both inside and outside of universities. The annual 'career guides' in the daily press are also singularly uninformed about the community's need for clinical chemists or even about the existence of these.

Technologists, technicians and other laboratory workers

Technologists

The Australian Institute of Medical Technology has succeeded in persuading various state institutes of technology which are similar to technical universities, but sometimes only of the status of a technical college, to start diploma courses in medical laboratory technology. In the past these were largely part-time courses, but now full-time three-year courses leading to a diploma in medical laboratory technology or degree in applied science are available.

Technicians

Besides the trained technologists, many laboratories are training their own technicians. Young people of school matriculation standard are appointed to the laboratories as trainees and are trained by personnel of the laboratory.

Some large laboratories arrange for lectures for technicians to be given by the senior graduate staff, but attendance of these lectures is not everywhere compulsory and rarely followed by examinations. The technical colleges have now instituted certificate courses by night lectures for technicians employed in medical laboratories.

The technologists in Australia are organized in the Australian Institute of Medical Technologists.

Possible developments

The major problems of clinical chemistry in Australia stem from the low population and vast distances. This is most notable in supplies and services.

Chemicals. These are largely imported from the UK, USA, Japan, but increasingly from Australian firms or subsidiaries. Supplies are not good in general; stocks held in most centers are meagre.

Instruments. Mainly imported. Service, in general, is not good and spares stocked by most importers are limited. Most laboratories have been severely handicapped by poor electronics and general maintenance arrangements.

Reference standards. All imported.

Control of equipment. Instrument suppliers have not made adequate provision for after-sales service in many instances. Laboratory staff have been forced to be self-reliant in many cases, which was acceptable (and probably highly educational) while instruments were not very sophisticated. With mechanization and highly sophisticated equipment however, lack of good maintenance services is a severe drawback.

The advent of the AACB has diminished the scientific isolation of many laboratories very significantly, but still most clinical chemists rarely are able to attend the national meetings. The problem of lack of specialization and centers of excellence is related to the lack of interest taken in clinical chemistry by universities in Australia.

Adequate professional training courses conducted by the AACB and by the universities will change the nature of clinical biochemistry in Australia during the next ten years. Larger installation of mechanized and automated analyzers and computers will enable a few more clinical chemists to move away from the common 'high demand' estimation and to develop highly-sophisticated specialized units of research and service. Clinical chemists will need to take care that these units are not all established in departments of medicine, surgery, cardiology, etc. If they are, clinical chemistry in Australia, as elsewhere, will face the bleak prospect of retaining only the highly repetitive 'common twenty' tests and losing its best brains and most interesting problems to the special units of other departments and research institutes.

In some Australian states there is and always has been a considerable degree of centralization which is more or less inevitable because of small and scattered centers of population. In the more populous states there has been up to now much more individualism in the laboratories of the larger public hospitals, individualism which has existed along with the more centralized services provided for limited sectors of the population by such agencies as the Commonwealth Health Department, the Repatriation Department and State Health Departments.

Some of these agencies will undoubtedly continue to provide laboratory services, though possibly with some changes in emphasis and scope as the public hospital systems expand. Surveys relating to administrative regionalization of health services are already proceeding and some pattern of hospital groupings is beginning to emerge. With the grouping of hospitals will undoubtedly come more rationalization instead of competition in clinical activities and more integration of laboratory services. Clinical chemists can expect to become increasingly concerned with this regionalization and integration.

National societies and journals

The Australian Association of Clinical Biochemists Inc.
This body represents clinical chemists in Australia. It has 400 members, most with science degrees and a few with medical qualifications. Most of those medically qualified are also members of the Royal College of Pathologists of Australia. The association is controlled by a council comprising one member elected by each state and a president, secretary and treasurer elected on a national basis.

Each state has its own branch and its own branch activities while there is one national scientific and business meeting each year.

The Royal College of Pathologists of Australia
Medically-qualified chemical pathologists are members of the RCPA. There is a great degree of harmony and co-operation between the AACB and the RCPA, with joint meetings, proficiency surveys, etc.

The RCPA publishes *Pathology* which as yet has contained few chemistry papers, although a member of the AACB is now an associate editor of the journal and more biochemical papers are expected to appear.

Most papers by clinical chemists in Australia have been published in medical journals, e.g. *Med.J.Aust.* or in clinical chemistry journals in other countries.

AUSTRIA

History
In Austria the first clinical chemistry tests were routinely performed in the Pathochemical Laboratory of the General Hospital in Vienna. This laboratory was founded in 1844 by Johann Florian Heller (1819-1871). Following a suggestion of the famous Viennese pathologist, Karl von Rokitansky (1804-1878), a Department of Medical Chemistry (Institut für Medizinische Chemie) was established at the School of Medicine, University of Vienna, in 1874. Its first chairman was Ernst Ludwig (1842-1915).

Legal status
The conduct of clinical chemistry is governed by the 'Verordnung des Bundesministeriums für soziale Verwaltung vom 2. April 1948 betre-

ffend die Befügnis zur Vornahme medizinisch-diagnostischer Untersuchungen und die hiebei bei Arbeiten mit Krankheitserregern zu
beobachtenden Vorsichtsmassnahmen' (Bundesgesetzblatt für die
Republik Österreich Nr. 63/1948).

In 1 of this 'Verordnung' is stated: 'Jeder Arzt ist befügt, zur Fest-
stellung einer Krankheit und zur Überwachung eines Krankheitsge-
schehens bei Kranken, die seiner Behandlung anvertraut sind,
chemische, physikalische oder mikroskopische Untersuchungen . . . .
selbst vorzunehmen, oder unter seiner Verantwortung durch fachlich
geschulte Hilfskräfte durchführen zu lassen'. (i.e. every practitioner is
allowed to perform diagnostic tests for his own patients without a special
training or a special license).

Hospital laboratories or non-hospital laboratories (private laboratories) can only be conducted by specialists for medical and chemical
laboratory diagnostics ('Facharzt für medizinische und chemische
Labordiagnostik') (cf. 3.2).

Education and training in clinical chemistry

Clinical chemistry in the medical curriculum

At the three Austrian medical schools (Vienna, Graz, Innsbruck) there
are at present no departments of clinical chemistry or clinical biochemistry. There are now three departments of medical chemistry in
Austria, one in each of the three medical schools. The departments of
medical chemistry are in charge of the training of medical students in
chemistry and biochemistry. In addition, there is a division of clinical
chemistry and laboratory medicine at one of the departments of
medicine at the University of Vienna.

Clinical chemistry, clinical biochemistry and laboratory medicine are
non-compulsory subjects in the medical curriculum. At the University of
Vienna a special course in clinical chemistry, including practical
exercises, is provided in the department of medical chemistry and in the
division of clinical chemistry and laboratory medicine at the department
of medicine.

Medical graduates

The 'Verordnung des Bundesministeriums für soziale Verwaltung vom
2. November 1969 über die Ausbildung zum Facharzt für nichtklinische
Medizin' (Bundesgesetzblatt für die Republik Österreich Nr. 450/1969)
introduced a new category of specialization that of 'Facharzt für nichtklinische Medizin' (specialist for non-medical medicine).

In 2 of this 'Verordnung' is stated: 'Als nichtklinische Sonderfächer
im Sinne Dieser Verordnung gelten:

(1) nichtklinische Lehrfacher die an der medizinischen Fakultät
einer Universität in der Republik Österreich vorgetragen werden;

(2) medizinische und chemische Labordiagnostik (i.e. medical and
chemical laboratory diagnostics);

(3) mikrobiologisch-serologische Labordiagnostik (i.e. microbiological-serological laboratory diagnostics)'.

In 3 of this 'Verordnung' it is stated that:—'Die Ausbildung hat zu umfassen:

(1) eine einjährige Ausbildung in einem oder mehreren klinischen Sonderfächern, davon mindestens sechs Monate im Sonderfach Innere Medizin;

(2) eine vierjährige Ausbildung im angestrebten nichtklinischen Sonderfach (cf. 2), sowie;

(3) eine einjährige Ausbildung in einem der nichtklinischen Sonderfächer (cf. 2) einschliesslich des angestrebten Sonderfaches.

The specialization in medical and chemical laboratory diagnostics include: one year practical training in medicine, at least six months in internal medicine, four years in medical-chemical laboratory diagnostics, one year in a non-clinical subject, including medical and chemical laboratory diagnostics.

Following this special training (six years), such medical doctors are recognized by the Österreichische Ärztekammer (Austrian Chamber of Physicians) as specialists in medical and chemical diagnostics ('Facharzt für medizinische und chemische Labordiagnostik'). Persons with such training are allowed to perform diagnostic laboratory tests in a responsible position (head of a diagnostic laboratory in a hospital or in a private laboratory).

Non-medical graduates
There is no definite curriculum for non-medical graduates in this country. Persons with a PhD degree in chemistry, biochemistry or biology can only work in a clinical laboratory under the supervision of a medical doctor who is a specialist in medical and chemical laboratory diagnostics.

Place of clinical chemistry in hospital service
In teaching hospitals attached to universities the large laboratories are usually part of the department of medicine. These laboratories are headed either by a medical doctor who is a specialist in medical and chemical laboratory diagnostics, or by a PhD directly responsible to the chairman of the respective department of medicine. Specialized laboratories in departments of obstetrics and gynaecology are part of these departments.

In other large hospitals (not attached to universities) there are, as a rule, central laboratories, serving all divisions of these hospitals. They should be headed by an MD specialist in medical and chemical laboratory diagnostics. In small hospitals, laboratories are usually under the direction of a pathologist, or are part of the division of medicine.

Place of clinical chemistry in non-hospital service
Private laboratories are under the direction of medical doctors who are specialists in medical and chemical laboratory diagnostics and have a special license issued by the Ministry of Health and Environmental Protection. Some of these laboratories are large establishments employing graduated chemists or biochemists. The majority however, rely on technicians to carry out their chemical work.

Technologists, technicians and other laboratory workers
The training of medical technicians is carried out in special schools for technicians (Schulen für die gehobenen medizinisch-technischen Dienste). There are now six such schools in this country (Graz, Innsbruck, Klagenfurt, Salzburg, Steyr, Vienna).

The curriculum for these studies is based on the 'Bundesgesetz vom 13. Feber 1969, mit dem das Bundesgesetz betreffend die Regelung des Krankenpflegedienstes, der medizinisch-technischen Dienste und der Sanitätshilfsdienste neuerlich abgeändert und ergänzt wird' (Bundesgesetztblatt für die Republik Österreich Nr. 95/1969).

The general requirements for entry to these schools is the examination qualifying for university entry ('Matura'). The curriculum for medical technicians includes the following subjects:

> Practical nursing (2 months)
> Anatomy and physiology
> General pathology
> Hygiene
> Inorganic and organic chemistry, biochemistry
> Histology and cytology
> Microbiology and serology
> Hematology
> Blood grouping methods (immunohematology)
> Photography, including microphotography
> Medical technology
> Medical documentation and medical calculations
> First aid
> Introduction to public health and social welfare
> Outlines of hospital administration

This course lasts two years and three months, during which time the participants attend lectures and gather practical experience in all the above subjects. After completion of this course, and on successful passing of a final examination a diploma is awarded.

Possible developments for the next ten years
In Austria the field of clinical chemistry is in a state of transition. In the future the introduction of more specialized biochemical analytical procedures and function tests will demand the organization of large

clinical chemical centers. These centers will be attached to big hospitals. Physicians, clinical pathologists, biochemists and analytical chemists will be working in such laboratory centers in close co-operation with the different departments of the hospital. Automation of analysis and data handling will increase the capacity of these laboratories and decrease the cost of the individual analysis.

Clinical chemistry laboratories in smaller regional hospitals will continue to carry out simple routine tests. Great care should be given, however, to the strengthening of contacts between these laboratories and large clinical chemistry centers. The staff of the clinical chemistry centers should regularly advise regional laboratories and should be exclusively responsible for specialized analyses which cannot be performed in small laboratories.

The future development of clinical chemistry will depend on the discovery and development of new analytical methods and on an improved training of clinical chemists. A clinical chemist in an up-to-date clinical chemistry laboratory should possess the following qualifications:

(1) sufficient technical and medical knowledge;
(2) knowledge of statistics and data handling;
(3) knowledge of principles and procedures of management;
(4) interest in research.

To meet all these needs departments of clinical chemistry or clinical biochemistry will be necessary at the three medical schools in Austria. The staff of these departments should primarily be devoted to applied research and development, as well as to the training of medical students. The department of clinical chemistry should also take care of postdoctoral training of medical doctors and biochemists holding a PhD degree, who wish to specialize in clinical chemistry.

National association and societies

An Austrian Society of Clinical Chemistry (Österreichische Gesellschaft für Klinische Chemie) was founded in 1968.

A Society of Medical Technicians (Verband der medizinisch-technischen Assistentinnen Österreichs) has been in existence since 1950.

BELGIUM

History

In Belgium, as in many other countries, the pharmacist was the pioneer of clinical chemistry because he had the laboratory of his office at his disposal and because his education prepared him to master analytical expertise. The first laboratory having general modern features was set up at the Military Hospital of Namur in 1912 at the initiative of W. Duliere, a military pharmacist.

However, the marked starting point towards an intensive development of clinical chemistry was given at the University of Liege (1920) by Professor Vivario (School of Pharmacy). He was officially charged by Royal Decree for the direction of a chemical research laboratory at the University Hospital. A course of clinical chemistry was made compulsory for the studies of pharmacy (Law 1929). No hospital may be accepted by the Ministry of Health if a laboratory is not available.

Legal status

With regard to regulations, clinical chemistry does not constitute a separate entity in Belgium. It is included as part of 'clinical biology', together with microbiology, serology, hematology, etc.

Legislation on clinical chemistry included in the legislation on clinical biology

Clinical biology is ruled by 'indirect' legislation. In fact anybody may set up a laboratory; however the Social Security Law establishes that expenses for analyses will only be refunded to persons having the analyses performed by those analysts certified as qualified by the Ministry of Health.

The analyst's qualifications

The Social Security Law prescribes that certification commissions at the Ministry of Health will establish the 'specialist' qualification for surgery, cardiology and other clinical disciplines. For clinical biology physicians and pharmacists are considered. There are two certification commissions, one for the physicians and one for the pharmacists. Qualification by a physician's or pharmacist's degree enables him to perform analyses appearing in the list of current medical care. To perform all the other analyses mentioned in the list of 'clinical biology' one has to be certified as a specialist in clinical biology. However, any specialized physician (for instance a gastroenterologist) may perform 'special technical performances' when they are related to the speciality for which they are certified. Thus, they may perform medical analyses.

The commission of physicians

This certifies any candidate who has proven that he dedicated four years of study, after having obtained his physician's degree, to clinical biology on a full-time basis in a university service or a service outside the university thus warranting his education and training. A degree exists for physicians specializing in biology. Most of the physicians are certified for all the branches except for pathological anatomy.

The commission of pharmacists

This has strict requirements as it certifies as pharmacists specializing in biology only candidates with university certificates for branches of clinical biology. Moreover, the regulation set up by this commission considers that the certification may be limited to only one branch and that can be withdrawn. The commission has the right at any time to have the laboratory checked by an Inspector of Pharmacy of the Ministry of Health.

Licencees will be obliged to provide proof of their participation in an official quality control program.

It should be noted that very few chemists or biochemists dedicated themselves to clinical biology. Their acquired rights were admitted only 20 years ago and they remained functional. Later other chemists or biochemists were not certified. A decree of September 1972 again authorizes certification, however this decree has been opposed by physicians specializing in biology who requested its suppression by the council of the state. This was done but a new decree of June 1975 makes certification now possible.

Education and training for directors and senior staff in clinical chemistry
No special study cycle or regulation is imposed in becoming the director of a clinical laboratory.

Place of clinical chemistry in hospital service
In the university hospitals there is an autonomous service of clinical chemistry. The responsible director is usually the professor teaching the subject at the school of medicine or pharmacy.

In the hospitals of large cities and private clinics, the clinical chemistry laboratory is a section of the clinical biology laboratory. The clinical chemistry activity reaches 40% to 75% of the total activity.

Place of clinical chemistry in non-hospital service
This type of laboratory is called a 'private laboratory'. Practically all of them belong to the 'polyvalent' type, and are under the jurisdiction of general practitioners or specialists.

Place of clinical chemistry in both medical and non-medical education
Scientific medical education, or more exactly education of biohuman sciences, is the concern of physicians and pharmacists. Scientific education without medical character depends on the faculty of sciences. For chemists and biochemists (faculty of sciences) some diversification is to be noted in the study program of the second cycle (bachelor of sciences). As a matter of fact the absence of any educational orientation towards clinical chemistry must be stated. With regard to medical and pharmaceutical education, the following characteristics appear.

Education for physicians
The teaching of essential human biological sciences such as cytology, histology, physiology and biochemistry is very developed in the first

cycle (candidature of medical sciences). This teaching is very thorough and useful as a base for the study of the pathological-biochemical aspect of clinical chemistry. During the second cycle (three years and one probation year for the degree of doctor in medicine, surgery and obstetrics) the teaching is essentially orientated towards diagnosis and therapy. The biochemical abnormalities caused by morbid affections are explained. It may be concluded that physicians are not well trained for the practice of clinical biology and consequently of clinical chemistry. This explains why, after obtaining the degree, four years of study are imposed for specialization.

Education for pharmacists

The program shows a submission on one hand to physics and chemistry and on the other hand to human biology. It is balanced with a very marked and very exacting orientation towards the analytical laboratory. This is the only study cycle in which clinical chemistry is legally compulsory (during the second cycle). It may therefore be concluded that the study program for pharmacy prepares the candidate for specialization in clinical biology and clinical chemistry. The education in pharmacy is characterized by a thorough teaching of bioanalytical expertise in the frame of human pathology. Therefore the duration of subsequent training of pharmacists is short.

Technologist and technician training

The technical education is quite variable.

Possible developments for the next ten years

Efforts will further be stressed on 'quality control' and standardized methods. To this effect the Société Belge de Chemie Clinique has sponsored the Société Belge pour le Perfectionnement des Méthodes, de Standards et des Unités officially registered as Nr. 6599 in the *Moniteur Belge*. Continued application of automation is expected for preventive medicine in large hospital units with judicious care in order not to depersonalize clinical biology to the disadvantage of the quality of good medical practice.

National societies, associations and journals

The Société Belge de Chimie Clinique was founded in 1958, and officially registered under Nr. 155 in the *Moniteur Belge*. In proportion to the size of the country, it has a relatively large membership. Taking into account the fact that Belgium is small and has two languages, there is no national journal. Dutch-speaking members publish in Dutch, German and English journals; French-speaking members issue their papers in French or English journals.

Every second year the Société Belge de Chimie Clinique organizes, in co-operation with the Société Belge des Sciences Pharmaceutiques, an International Symposium on electrophoresis and chromatography; our association assumes responsibility for the biology section.

Economic status of clinical chemists
This is variable and not legally regulated.

BOLIVIA

History and background
In Bolivia the title of clinical chemist and clinical biochemist does not exist. Those with valid degrees and titles are the practising chemists and biochemists. Schools for pharmacists have only existed since 1930-1935, and pharmacy training enabled students to obtain the title of pharmacists and chemists. Later, increased knowledge of advanced medical sciences was provided by those who had studied in foreign universities in clinical laboratories. With this improved medical knowledge, training programs were established which fulfilled the needs for proper training in the specialities of biological sciences and pharmacy.

Legal aspects
According to Bolivia's present national state laws, a practising clinical chemist can only function for the professional biochemists. The national regulations for laboratories and pharmacists were established by the government as follows.

(1)   The foreign degrees of clinical pathologists or medical technologists and physicians are not recognized by our laws unless they are authorized by the OMS (World Health Organization).

(2)   The maximum authority of public health is vested in the national director of laboratories and pharmacists. Each state has inspectors to examine laboratories.

(3)   Biochemists and pharmacists must be nationally certified. State certification is optional.

Societies
The main professional society of Bolivia was recognized by the government and named The National School of Biochemistry and Pharmacy It is a member of the Latin American Society of Clinical Biochemistry, and plans to become a member of the IFCC.

Education
In the University of Bolivia, the professions of pharmacy and biochemistry have been separated (reform was initiated in 1972). Previously these professions constituted one career, accounting for the name of the institution, The College of Biochemistry and Pharmacy.

The profession of biochemist allows a person to work in all the specialities of a clinical laboratory.

All the laws mentioned before are valid for this kind of professional speciality. There are three colleges in the University of Bolivia that teach biochemistry. All of them are similarly orientated and programmed by the National Council of Superior Education. This is empowered to direct and orientate all the universities of the country. Clinical biochemists from Bolivia and other Latin American countries are considered to have the same academic and legal status as those of other countries.

Additional considerations

The development of clinical laboratories in Bolivia is well under way. In some urban areas some private and government laboratories are more or less well equipped. However, there are still many regions without clinical laboratory services. The expansion of medical specialization and the clinical laboratories is growing at the same rate. In rural areas the government agencies are working very actively, but on a small scale, to get the benefit of this work to the poor people in remote areas. Although the universities of biochemistry and other institutes with international help, for example, the Institute of Biology of Altura are very active research centers, they are still in the process of development. Additional international help and specialized libraries are also needed. The Institute of Bacteriology of La Paz, is contributing very effectively to public health and research.

BRAZIL

History

By definition, clinical chemistry is the art of biochemical analyses for clinical purposes as an aid to the medical profession. By law, the laboratories for clinical analyses and research, private or public, separate or within hospitals, may function only under the responsibility and technical direction of officially licensed pharmacists or physicians. The former, who are required to have undergone specialized training courses in the whole field of clinical analyses, including clinical microbiology, immunology, virology, cytology, biochemistry, hematology, parasitology, toxicology and general pathology, are now designated as 'pharmacists-biochemists'. For physicians there are no requirements or opportunities for specialization in medical schools up to now. Any person with a university degree of medicine or pharmaceutical biochemistry may open a laboratory for clinical analyses under a license issued by the National Department of Public Health, the license has to be renewed every year. However, there are plans for future restrictions and more rigid requirements with the obligation for further specialization in the field. There exists no category of clinical chemists or clinical pathologists as such.

At the present time, there are two representative private professional societies: the Sociedade Brasileira de Análises Clínicas—SBAC, and the Sociedade Brasileira de Patologia Clínica—SBPC, both of which issue diplomas for clinical analysts or clinical pathologists, respectively, to members who must pass a special examination. The SBAC is open to all professionals interested in working in clinical analyses, whereas the SBPC admits only physicians as effective members.

Although pharmaceutical sciences have been taught in Brazil since 1832, clinical chemistry has appeared only recently as a separate subject on the curriculum of pharmaceutical schools. Courses of medicine teach only clinical pathology placing more emphasis on the interpretation of clinical data than on technical details for the execution of clinical analyses. Pharmacy students who have completed their instruction courses in basic disciplines are offered the choice of specializing in either commercial pharmacy, public health, or technology. In the last years, more than 80% of these students have selected the public health sector which includes clinical analyses.

There are now 26 teaching establishments, 15 of which bear the name of faculties, 21 of these schools are maintained by the federal government, the remaining five by the states. One state has four faculties, two have three, and the rest have two or one, with no teaching facilities in seven states.

There is one pharmacist to 5,500 inhabitants and to 500 square kilometers, summing up to 18,000 professionals, although most of these are not working in clinical analyses.

There are about 2,500 laboratories for clinical analyses in the country, the majority in the large population centers. Most of these are managed by pharmacists, especially in smaller towns and the interior of Brazil.

The number of students matriculated in the health area, including nursing schools, medicine, dentistry and pharmacy, has now reached 60,000 about 11% of whom are studying at schools of pharmacy. Nearly 60% of these are women. About 1,200 pharmacy students are expected to graduate as specialists in clinical analyses by the end of the year.

The teaching staff comprises nearly 700 professors of pharmacy and biochemistry. There are five postgraduate courses, one of which is dedicated to clinical analyses.

Although there are some frictions and incompatibilities of views between physicians and pharmacists, both professions are working harmoniously together for the improvement and preservation of public health in Brazil, which faces many problems of lack of resources, difficulties of communications, dependence on imported scientific material and unusually hard working conditions in this very extended tropical region which is continually menaced by epidemics, natural catastrophies and low economical prospects, which hamper but at the same time encourage the socialization of medical treatment including the field of clinical chemistry.

BULGARIA

History
Historically, clinical chemistry has evolved from the needs of clinical practice. Owing to the spontaneous development of clinical chemistry its meaning, content and tasks have not been clearly outlined. This holds good especially for the demarcation of clinical chemistry from the remaining branches of biochemistry. In this respect one comes across differences in the opinion of various clinical chemists.

Classification of clinical chemistry
In Bulgaria the development of clinical chemistry has been based upon the following classification of the branches of biochemistry.

    (a)   Biochemistry of Plants
    (b)   Biochemistry of Animals
    (c)   Biochemistry of Man

    (1)   Normal biochemistry (orthobiochemistry)
    (2)   Pathobiochemistry ⎫
                              ⎬   Clinical biochemistry
    (3)   Clinical chemistry ⎭
    (4)   Pharmacological biochemistry

Biochemistry of man, which is related to human medicine is subdivided in Bulgaria into normal biochemistry (orthobiochemistry), pathobiochemistry, clinical chemistry and pharmacological biochemistry. Normal biochemistry deals with the mechanisms of the normal metabolic processes. Pathobiochemistry deals with the mechanisms of pathologically altered metabolic processes. Clinical chemistry embodies the analytical chemical methods of studying both fluids and tissues for clinical (diagnostic, prognostic and therapeutic) purposes and the clinical interpretation of the results obtained. Pharmacological biochemistry (or biochemical pharmacology) deals with the mechanisms involved in the effect of various drugs on normal and pathologically disturbed metabolism.

In accordance with the above definitions special emphasis is laid on the differences between clinical chemistry and pathobiochemistry, maintaining that these are two distinct branches of biochemistry so that there is ground for identifying them. Pathobiochemistry represents chemical pathophysiology, whereas clinical chemistry represents chemical diagnosis. The combination of clinical chemistry and pathobiochemistry is termed clinical biochemistry. In this subdivision it is not just the names that make the point, but the content they imply.

At present, pathobiochemistry is regarded as a subdivision of clinical chemistry. However, pathobiochemistry is a theoretical discipline which may exist only in research institutes and in universities. This is not in accordance with the activities of the majority of the clinical chemists, who are involved in routine hospital practice. Hence the separation of pathobiochemistry from clinical chemistry is urged for practical reasons as well.

At the Postgraduate Medical Institute in Sofia, for 12 years there have been two separate departments of clinical chemistry and of patho-biochemistry. The latter will evolve in the coming years into an independent chair of pathobiochemistry. The postgraduate training of physicians is provided by two separate sets of lectures in clinical chemistry and pathobiochemistry, with separate textbooks for both disciplines. Both departments differ in their scientific problems; the chemical diagnostic tests are carried out by the clinical chemistry department.

CANADA

History
Clinical chemistry in Canada has developed rapidly as a vigorous speciality since the formation of a national society, The Canadian Society of Clinical Chemists (CSCC) in 1956. Prior to this period, there was no organization which represented the specific interests or met the specific needs of clinical chemists. As analytical methods came into use in the hospital laboratories in the early part of this century, chemistry became a section of clinical pathology and some basic biochemists were attracted into the hospital chemistry laboratory. With the explosive development of analytical clinical chemistry since the 1950s, it became apparent that clinical chemists in greater numbers were required to take responsibility for the increasingly complex chemical operations in the modern hospital laboratory. During its short history the CSCC has advanced the clinical chemist to a position of influence in the medical community.

A number of training programs now exist and others are being developed in Canada for both medically and non-medically qualified clinical chemists at postgraduate and postdoctoral levels. With the establishment by the CSCC of a certification program and the supporting educational programs, continued improvement of the professional status of clinical chemists, and elevation of the standards of practise of clinical chemistry in Canada is assured.

Legal status
In Canada, health is a provincial matter. In most provinces there exists, at the present time, little or no government legislation concerning the practise of clinical chemistry. This matter is now being explored in some provinces by provincial medical associations. In the province of Ontario,

however, all clinical laboratories are now controlled by government legislation. Clinical chemistry laboratories (hospital and commercial) must be licensed and the license holder must have qualifications defined by government in order to practise. (MD or PhD with experience, technologists). Standards are maintained through onsite inspections by government inspectors and an extensive laboratory proficiency testing program. It is expected that similar measures will be instituted by all provinces in the near future.

Organization of clinical chemistry laboratories

Hospitals

The hospital clinical chemistry laboratory may be a division of a department of laboratories, or a department of medicine, or a department of pathology. It may also operate as a separate, semi-autonomous department of clinical chemistry, and this is the trend in large university hospitals. Although the operation of the routine laboratory is his most important function, the clinical chemist is expected to be able to discuss the nature and implications of biochemical disorders and the chemical interpretation of laboratory results with the medical staff.

The teaching of clinical chemistry or biochemistry to the hospital staff, residents and interns, technicians, nurses, through formal lecture courses and presentation of data and discussions at teaching ward rounds of the hospital is a responsibility of the clinical chemist. In university hospitals, senior clinical chemists, usually hold an academic appointment and may direct graduate student's research and contribute to the graduate and medical undergraduate teaching program of the university.

Government laboratories

The provincial governments of Canada operate centralized laboratories which are organized usually into separate divisions within a 'department of laboratories', directed by a medically-qualified laboratory scientist. Some of these laboratories have large modern facilities for clinical chemistry, and provide regional services to hospitals, doctors in practice and various studies in the health field.

In the Department of National Health and Welfare in Ottawa (Federal) the Laboratory Center for Disease Control, serves the general public, hospitals and doctors on a country-wide scale. Its clinical chemistry laboratory has developed a standardized procedure for hemoglobin measurement across Canada. It has also undertaken a program of evaluation of clinical chemistry kits, and instrumentation. The results of these evaluation studies are published regularly in a *Newsletter,* The Central Laboratory in Ottawa has also published for some years a *Manual of Clinical Chemistry Procedures for Hospital Laboratories.*

Commercial laboratories

In the past few years there has been a very rapid growth of commercial clinical laboratories in Canada. These service laboratories range from smaller laboratories, which provide only a few basic clinical chemistry tests and serve doctors' clinics, and doctors in private practice, to large centralized laboratories which provide specialized clinical chemistry services.

Education and training for directors and senior staff of clinical chemistry laboratories

The Canadian Society of Clinical Chemists has instituted a program of examinations for certification of clinical chemists. This program is open to non-medical and medical graduates who wish to become certified in the CSCC. The Royal College of Physicians and Surgeons in Canada has established a fellowship in medical biochemistry which is available to medical graduates only. Training programs are being actively developed in Canada to achieve the goals defined. These are now available at a number of Canadian universities and hospitals, and new programs are in various stages of development. Continuing education programs also exist and developments in this area are of major concern.

Medical graduates

The medical biochemistry fellowship program is a five-year program in which two of the five years must be spent in a structured program of training in clinical chemistry in an approved hospital. A number of universities and hospitals have established, or are planning courses which meet this requirement.

At the University of Montreal and the University of Alberta, there in clinical chemistry has been established since 1968. The course is available to graduates in medicine who have an honours science background. The objective of the diploma course is to develop a satisfactory level of competence in those responsible for directing a clinical chemistry service laboratory in a hospital. It provides intensive training in clinical chemistry as a practical, as well as theoretical level. Courses cover the biochemistry of disease processes; immunological and radio-isotope techniques; biometrics; analytical biochemistry and laboratory administration. Candidates are assigned to the clinical chemistry laboratory of an approved hospital service laboratory. The diploma is awarded on the basis of a written examination (CSCC) on completion of the two-year program. After an additional year of experience, diploma graduates are eligible to sit the certification oral examinations conducted by the CSCC.

The department of pathological chemistry at the University of Western Ontario has become a division of clinical biochemistry in the department of biochemistry. Postgraduate programs are being developed for medical graduates who are proceeding to a Royal College fellowship in medical biochemistry.

In Quebec, McGill University offers a course for medical graduates which is administered by a sub-committee for residency training in medical biochemistry. This is a five-year program, in which candidates are required to complete a one-year internship, three years in an approved laboratory and one year in the clinical chemistry laboratory of a teaching hospital. Suitable candidates may concurrently undertake a PhD program in the departments of biochemistry, physiology, pharmacology or experimental medicine. A similar program leading to a DSc in medical biochemistry is available at Laval University.

At McMaster University, a division of clinical chemistry within pathology offers a five-year residency program to prepare medical graduates for the Royal College fellowship examinations in medical biochemistry.

In other parts of Canada, similar training programs in medical biochemistry are in various stages of development.

Non-medical graduates
A few training programs in clinical biochemistry have now been established in Canada, which are available to both non-medical and medical graduates. The program at the University of Toronto recognized by the CSCC as suitable preparation for its certification program, is offered as a two-year postdoctoral diploma course in clinical chemistry to candidates with the PhD or MD degree. A program is also available at this center leading to a PhD in clinical biochemistry.

Graduate programs in clinical chemistry have been established at the University of Windsor leading to an MSc and PhD in clinical chemistry. The programs are directed by the department of chemistry in co-operation with clinical pathologists who direct the clinical chemistry service laboratories in Windsor hospitals.

At the University of Montreal and the University of Alberta, there have been established four-year undergraduate programs leading to the BSc degree in clinical chemistry. These programs provide excellent training for individuals who wish to proceed to more advanced qualifications. Graduate and postgraduate courses in clinical chemistry are planned to begin in 1973.

In British Columbia, postgraduate programs for clinical chemists have been established at the Royal Columbian Hospital, New Westminster and the Vancouver General Hospital. Candidates spend two years in a structured program in the hospitals' clinical chemistry laboratories.

At McMaster University, a newly proposed program for postdoctoral training in clinical chemistry is expected to get underway within the next year. A three-year program has been planned to PhD candidates, which is designed to meet the requirements for certification established by the Canadian Society of Clinical Chemists.

Additional programs for non-medical graduates, at graduate, postgraduate and postdoctoral levels, are presently under consideration at a number of other Canadian universities.

Education for Clinical Chemistry

The Education and Certification Committees of the CSCC have established a program for accreditation of all training programs offered at the postgraduate and postdoctoral level. Graduates of accredited programs may proceed toward certification by the CSCC.

Clinical chemistry in medical education

In Canada a few medical schools have a regular course in clinical chemistry for undergraduate medical students. In others, the subject is fragmented in a systems oriented medical curriculum. Medically qualified staff tend to predominate in teaching clinical chemistry to medical students. The professional clinical chemists in the university hospitals participate in clinical pathological conferences and special presentations at general and speciality rounds. The need for more vigorous training in clinical chemistry in the undergraduate medical curriculum is recognized in some centers and it is probable that medical biochemists and clinical chemists will become more involved in the teaching.

Clinical chemistry in non-medical education

In several universities clinical chemistry is still available as a fourth-year option in undergraduate chemistry and biochemistry programs, and it can be the major subject of MSc and PhD programs. There are also postdoctoral training programs in clinical chemistry in a few centers. These educational programs are directed toward preparing candidates for certification in the CSCC as professional clinical chemists.

In medical technology training programs, clinical chemistry is taught at the basic level, and advanced courses are available for technologists seeking higher qualification in the Canadian Society of Medical Technologists. The professional clinical chemists actively assist in the development of basic curricula and the organization and presentation of advanced courses.

Technologists, technicians and other laboratory workers

In Canada the medical technologists are officially represented by the Canadian Society of Laboratory Technologists (CSLT), which through its national registry has established the Registered Technologist (RT) as the standard of qualification for the practice of medical technology. Technologists obtain their training only in training schools approved by the Canadian Medical Association's Committee on Approval of Training Programs for Medical Laboratory Technologists. A didactic eight to eleven months' program is given in colleges, central training schools, institutes of technology, departments of public health and larger hospitals followed by one year of supervised apprenticeship training in an approved hospital. Advanced courses of training leading to higher technical qualifications, e.g. ART and Licentiate, are available.

In addition to these programs, some Canadian universities have for some years offered courses in medical technology leading to the BSc degree.

E.T.C.C.—C

Individuals without formal medical technology qualifications, who are chemistry graduates from universities or technical schools, are often employed in the clinical chemistry laboratory for special tasks, (e.g. method development projects). One may find, also, 'technicians' without formal training occupied in positions of a more clerical nature.

National societies, associations and journals

The Canadian Society of Clinical Chemists is a national body which represents the interests of the clinical chemists in Canada. Since its formation in 1956, the society has grown to a present membership of approximately 350, with members in every province of Canada. Full membership in the CSCC is held by persons mainly engaged in the practice or teaching of clinical chemistry who have been so engaged for at least two years and who are graduates in science or medicine. Associate members are persons with an interest in clinical chemistry, but not possessing one or both of the above qualifications. The CSCC has established standing committees on methods, instrumentation, education, quality control, certification, and professional relations.

A national meeting is held annually and on a number of occasions it has been held jointly with the American Association of Clinical Chemists (AACC). In 1975, the CSCC and the AACC undertook the organization of the Ninth International Congress on Clinical Chemistry in Toronto, Canada.

The Canadian Association of Medical Biochemists was formed recently to represent the medical biochemists. Membership is limited to individuals having the MD degree. This national body is concerned with the development of training programs for MDs seeking the Royal College of Physicians and Surgeons fellowship in medical biochemistry, and the development of continuing education programs for its members.

Provincial organizations

As health is a provincial matter in Canada, provincial societies are now being formed. Four provinces, Ontario, British Columbia, Nova Scotia and Manitoba have now established provincial societies of clinical chemistry. The provincial societies are sanctioned by the CSCC and operate under the regulations laid down by its Constitution. The provincial societies will represent the interests of the clinical chemists on all matters relating to government—salaries, laboratory accreditation programs, etc.

The province of Quebec has a well-developed provincial organization which operates independently of the CSCC. Biochemists working in hospitals in the province of Quebec formed the Quebec Hospital Biochemists Corporation in 1961. The principal aims of this corporation are to promote, and develop the professional interests of clinical biochemists in Quebec. Its activities are published in its official *Bulletin* three times yearly.

Journals

In June 1968, a new international journal—*Clinical Biochemistry*—was introduced as the official journal of the Canadian Society of Clinical Chemists. The society also publishes a bi-monthly *Newsletter*.

The *Bulletin de la Corporation des Biochemistes des Hopitaux du Quebec* is the official journal of the Quebec Hospital Biochemists Corporation.

Future developments of clinical chemistry

The future of clinical chemistry in Canada will depend on the efforts of clinical chemists to foster and expand the programs of training and education which have only recently been established. These programs have been designed to produce clinical chemists and medical bio-chemists at a level of competence that is now defined by certifying bodies. Only through these activities will clinical chemists achieve recognition, acceptance and equality as full members of the medical community. It is likely that over the next five to ten years, the academic aspects of clinical chemistry will become more closely co-ordinated with the service aspects. Indeed, several university centers envisage the setting up of university departments of clinical biochemistry in the teaching hospitals, with the academic and service functions centralized in one department. These departments would be staffed by an appropriate blend of medical and non-medically qualified clinical biochemists having responsibilities in varying degrees for academic teaching, research, applied research, and service. It is believed that in this climate, clinical biochemistry will flourish and the most effective training programs can be developed.

The increasing costs of providing health care will inevitably lead to increasing regionalization and co-ordination of laboratory services. Indeed, co-operative shared resource facilities already exist in some parts of Canada where specialized services are located in closely grouped hospitals, each hospital providing services to the others in the group according to local resources of equipment and expertise.

It is probable that in the near future government legislation will be enacted in all provinces in Canada, concerning licensing and inspection of all clinical laboratories as well as of the individuals who take responsibility for the services provided.

CHILE

History

For more than forty years many names have been used in Chile for what is today known as clinical chemistry. At the beginning it was called chemical pathology, biological chemistry or medical chemistry, later it became pathological chemistry, diagnostic biochemistry or clinical biochemistry. It will be difficult for the term 'clinical chemistry' to be introduced in this country because the biochemistry and clinical pathology departments of the different universities are responsible for the teaching of this medical science. Clinical biochemistry is the more popular term in Chile.

In general in Chile, the name 'pathologist' is given to a medical doctor specialized in morbid anatomy and (or) histology. The name 'laboratorist' is given to a medical doctor, biochemist, chemist or medical technologist specialized in clinical chemistry, hematology or microbiology.

Legal status

As long as there have been medical doctors in Chile, the speciality of clinical laboratories has legally existed. Today there is a growing body of laws that control the work and duties of public and private laboratory medical doctors, clinical biochemists, chemists and medical technologists. The director of a clinical laboratory in Chile according to the present law can be a medical doctor, a biochemist, a chemist with pharmaceutical education or a medical technologist in that order and according to available manpower, location and size of the hospital. In the private sector any of these professional individuals can become the director.

Societies

Three professional societies exist today in the country, all of them legally recognized by the government.

[1] Chilean Society of Laboratory Doctors

Created in 1965, it includes 80 to 90 medical doctors with the speciality of clinical laboratory science. Since 1965 the society publishes a journal of small circulation [*Lab. Clin. Chile*]. Monthly scientific meetings are also arranged, exclusively for members and special guests.

[2] Society of Laboratory Biochemists and Chemists

Organized in 1944, it includes about 150 chemistry and 50 to 60 biochemistry graduates. The members are high-level professionals appointed in universities, government agencies and National Health Service hospitals. The society is divided into several local sections and is very active in scientific aspects; is affiliated to the Latin American Association of Clinical Biochemistry and to the International Federation of Clinical Chemistry.

[3] College of Medical Technologists of Chile

Established officially in August 1969, it existed previously for a number of years. A membership of 1,032 medical technologists was enrolled in 1973, distributed as follows in the different specialities:

Clinical laboratory	566
Hematology and blood bank	164
Histology and pathology	116
Radiology	109
Ophthalmology	70

The college has become very active in the last few years organizing meetings and graduate courses.

Clinical chemistry in the university

Status

Clinical chemistry departments or institutes are rare in Chile. One can find biochemistry departments with a small group of people devoted to clinical aspects, pathology departments with groups investigating clinical biochemistry problems, or biological chemistry departments with basic rather than clinical orientation, linked to every one of the nine existing medical schools, in the country. Besides, and belonging to other faculties, there are two or three departments or institutes of pathological chemistry and clinical biochemistry.

Many of these institutions are rather well equipped and staffed but with very little contact with the routine clinical laboratories. There is a need for the organization of departments at university level devoted solely to clinical chemistry problems with a strong link to the routine laboratories. The reinforcement of library facilities in this field is also an acute need.

Medical education

In the nine existing medical schools the teaching of clinical chemistry is incomplete. The students take basic biochemistry and physio-pathology during the first three years with some topics in clinical biochemistry and little experience in routine laboratory work. During the internship period (fifth to seventh year) they have a short course on emergency procedures and urinalysis. Those few interested in the laboratory can specialize by taking a three-year postdoctoral fellowship offered by the National Health Service. These programs however are unattractive and ill-organized, with the result of a declining interest for young doctors to enter the laboratory speciality.

The Austral University of Chile has now started a formal clinical biochemistry course at the end of the third year for medical students. A laboratory residency is also under study. Similar efforts are under study in the other medical schools.

Biochemists and chemists

Through the faculties of sciences and chemical sciences, two universities in Chile offer the degree of biochemist and chemist. Students with chemical-pharmaceutical education are also prepared in clinical chemistry problems. The master and doctor degree programs however, do not offer specific training in clinical chemistry; interested individuals must seek their graduate training abroad.

Medical technologists

Since 1949 the medical technology schools have been affiliated to the universities. Today there are six such schools in various parts of the country. The studies include three years of basic sciences and one year of guided-practice. They provide an excellent medium-level professional qualification for the laboratory, enabling techniques and procedures to be carried on independently.

Clinical chemistry in hospital services

The health care of about ten million inhabitants in Chile is provided by three groups of hospitals:

> University teaching hospitals
> National Health Service hospitals
> Private hospitals and clinics

There are nine university hospitals in the country and each one is equipped with rather modern clinical laboratories divided into chemistry, hematology, bacteriology, histology and parasitology sections. The heads of these laboratories are usually medical doctors and they are staffed with doctors, biochemists, chemists and medical technologists. They do research according to their capacities and provide routine services for the hospital. Spots of very high sophistication can be found but the link with the basic departments is weak.

The National Health Service (NHS) is by far the largest hospital owner in the country (232). These are divided into three categories: regional centers (12), intermediate and rural hospitals. Here, the shortage of manpower and laboratory facilities is evidenced in all its reality.

The 'regional laboratories' are in many cases short of staff, equipped with manual apparatus and ill-organized and administered. They are isolated from each other and from the center of the country with poor technical service, supplies and budget. None of them is engaged in research nor development projects in clinical chemistry.

The 'intermediate laboratories' located in small cities are in a more deteriorated situation in terms of staff, space, equipment, supplies and budget. They are usually of the size of one or two rooms with a chemist and (or) a medical technologist in charge and with three to four laboratory assistants at the most. The methods carried out include: blood sugar, BUN, urinalysis, blood counts, blood grouping and cross-matching. In some cases total protein, cholesterol, creatinine and some enzymes are also performed.

'Rural laboratories' are not existent in many cases or not staffed in others, the rest of them are poorly set up and directed by a medical technologist often working alone and isolated.

There are a certain number of small 'private hospitals', most of them located in the capital city. They take care of a small but sizeable segment of the population that can afford private services. Their clinical laboratories reflect the level of those in provincial centers but are business-orientated rather than service and development orientated. Similar comments can be said for private laboratories in non-hospital services. They constitute a minute part of the total work done in clinical analysis in the country.

Clinical chemistry: present and future

The clinical chemistry gap between developed and developing countries is widening at an astonishing rate. Even though everybody recognizes the impact of this new science in modern medicine, developing nations, mainly because of economical reasons, have been unable to cope with the accelerated development in this field. Chile is not an exception. It has essentially the same problems as are prevelant elsewhere in Latin America.

At the moment Chile does not have a National Reference Laboratory in clinical chemistry and the different laboratories work in isolation. In 1928 the government created the Bacteriological Institute of Chile, affiliated later to the NHS, with the mandate of organizing and recommending guidelines for all the aspects of the clinical laboratory through the organization of different sections, specially in bacteriology. However, because of low priorities and lack of funds the few guidelines proposed have not been implemented. Programs in bacteriology and tuberculosis are now in progress with fair results considering the economical tribulations and the difficult geography of the country.

It is hoped that the government will start a project for the rationalization and reorganization of laboratories services at national level, and a functioning advisory body with participation of the NHS, universities and professional laboratory societies should be organized. There should be laboratories with one national and several regional interconnected centers which would help the decentralization of decisions. A clarification of priorities, goals and available funds will also be beneficial. The substrate exists; nowadays in Chile there is a certain amount of young, enthusiastic and qualified manpower, and many of them have the will to help in the improvement of the status and services in clinical chemistry.

COLOMBIA

History

The speciality of clinical laboratory studies has legally existed for more than 50 years. Many names have been used for workers in this speciality: microbiologist and clinical laboratorist; bacteriologist; clinical laboratorist; medical technologist; medical technician; these names apply to people who hold a degree other than medical doctor (MD). When a medical doctor is trained in clinical laboratory science, usually, he or she is called a 'medical microbiologist' or 'medical bacteriologist'. In the last ten years the speciality of clinical and anatomical pathologist has been developed and includes medical doctors with training in clinical laboratory science and anatomical pathology.

The term 'clinical chemistry' has not been introduced in the country, because the clinical laboratory training is conducted by the schools of laboratories or departments of pathology.

Legal status
There are several governmental regulations concerned with the work of medical technologists, clinical laboratories, and clinical pathologists. The director of a clinical laboratory must be a medical doctor or micro-biologist or clinical laboratorist. All persons who have received training in clinical laboratory science (non MDs), after graduation, must work for one year in a public health clinical laboratory in order to obtain a license to practise the profession.

Societies
There are four types of professional societies:

[1] *SOCOPAT:* [*Colombian Society of Pathologists*]
This includes about 130 doctors with the speciality of pathology, mostly anatomical pathologists. Every year there is a national meeting with the presentation of several scientific papers.

[2] *ASBAS:* [*Association of Bacteriologists and Clinical Laboratorists*]
This society was organized about 12 months ago and includes around 260 members, mainly bacteriologists and clinical laboratorists. There is an annual meeting and scientific conferences take place throughout the year in the several local sections. The association works also as a labor union for their members helping them in relationship with the different institutions where professionals are appointed.

[3] *ALCE:* [*Association of Specialized Clinical Laboratorists*]
This society includes a group of professionals who work together trying to improve the facilities of the laboratories and the quantity and quality of laboratory tests. Since 1964 the society has published a journal called *Bulletin of Clinical Laboratories.*

There is a national scientific meeting every year where members and special foreign guest speakers participate.

[4] *Colombian Society of Microbiologists*
Similar to ALCE with which they work closely. The journal is published by the two societies.

Education
There are several universities with training in clinical laboratory studies. The length of the course varies between three and four years and the varying titles of the course create some confusion. The government is negotiating agreements between the universities and schools of clinical laboratories in order to clarify this situation and to establish only one title, probably that of clinical laboratorist course. In the case of clinical pathologists, a medical degree is required, followed by three years of a clinical and anatomical pathology residence program.

In general, all the programs for clinical laboratory studies are well developed, with good facilities and training in the different fields such as: hematology and blood banking, microbiology, clinical chemistry, immunology, parasitology, advanced microbiology and biochemistry, cytology and histology.

Types of clinical laboratories
There are several big clinical laboratories located in university and regional hospitals. These hospitals are affiliated with universities and the clinical laboratory students receive practical training in those laboratories. In general, they have good facilities and well-qualified staff.

There are several private laboratories in most of the cities, some of them well organized with good equipment and well-qualified staff. In small cities or communities some laboratories exist including the public health centers and small private laboratories.

As was mentioned before, the profession of clinical chemistry does not exist in Colombia as an individual discipline.

CZECHOSLOVAKIA

Legal status
In Czechoslovakia only a physician may serve as director of the laboratory.

Education and training for directors and senior staff of clinical chemistry laboratories
The postgraduate education from the completion of medical school until the end of the doctor's career would fall into two stages; in the first (the specialization system), the graduate with a general knowledge of medicine should be trained in one of the 19 basic specialities.

The young doctor enters a hospital for a compulsory practical course during which attention is mainly focused on internal medicine and surgery. Training for the selected speciality continues for the next three years, leading to a first-grade specialist examination. First-grade specialists can continue their work wherever such a qualification is adequate for their duties, e.g. in polyclinics, or they can go on with their training in a specialized department with a view to obtaining a second-grade specialist qualification.

Some of them select a narrower discipline, e.g. clinical biochemistry, aiming to become highly specialized and continue their training in an appropriate department of clinical biochemistry. Training for such a speciality continues for a further three years.

The second-grade qualification in a highly specialized discipline, e.g. clinical biochemistry, is one of the conditions for obtaining a senior post, head of the department of clinical chemistry or his deputy.

Clinical biochemistry in Czechoslovakia is established as a medical profession. Only a physician specializing in clinical biochemistry is allowed to manage a hospital-department of clinical biochemistry. Other graduate chemists, non-medical by profession, are employed in these clinical biochemistry departments too but not in the position of heads, and they do not participate in speciality qualification training.

The second stage in postgraduate training is further improvement of the specialist's knowledge which is systematically carried out while he is fully engaged in his job.

For all postgraduate studies and specialization of physicians in Czechoslovakia the Postgraduate Medical Institute, an establishment of the Ministry of Health, is responsible. The chair of clinical biochemistry of this institute manages and performs education of those physicians who decide to specialize in clinical chemistry. This specialization lasts six years, the first three years being devoted both to clinical bed-side duties in departments of internal medicine and to work in the bio-chemical laboratory. In the next three years the candidate works mainly in biochemical laboratories. During this time he can attend special courses or gain a few weeks' scholarship in the chair.

The period of education is then completed by a special examination given by the chair of clinical chemistry.

Place of clinical chemistry in hospital and non-hospital services

In addition to managing the clinical chemistry laboratory, the main duty of the head of the chair of clinical chemistry is to perform cyclic advanced courses for heads of clinical chemistry departments as well as thematic education for all specialists, both physicians and chemists interested in concerned problems.

The members of the chair are physicians and chemists and give lectures in special problems of clinical chemistry courses, organized by other medical chairs of the institute.

Place of clinical chemistry in medical education: undergraduate

Medical undergraduate education in Czechoslovakia is provided by nine medical faculties. The course lasts six years.

The first years at all faculties have a common curriculum consisting of theoretical subjects: biology, chemistry and biochemistry, physics, anatomy and physiology. In the third year the curriculum covers preclinical subjects: pharmacology, pathology and clinical subjects. The fourth and fifth years are devoted to clinical medicine and the sixth year mainly to a more systematic practical preparation for the final examinations.

Place of clinical chemistry in medical education: postgraduate

Institutes for postgraduate medical education
The purpose of postgraduate training is not only to attain a higher level in the doctor's chosen speciality, but also to consolidate and accord his knowledge with the progress and development of sciences. This is a high purpose which can be attained in a variety of ways: by reading specialist literature, by attending lectures and seminars at the place of work, by participating in congresses and conferences of scientific societies, by practical and outside assignments to other institutes and so on. These are by now generally accepted and undoubtedly effective methods.

However, all the above-mentioned approaches are fragmentary, lacking in system and do not always assure acquisition of a general grasp of the disciplines concerned. Their use depends in a considerable degree on the personal initiative of the doctor, the time he makes for it, and many other factors. The importance, indeed, the necessity of organizing systematic postgraduate medical teaching is, therefore, now generally accepted.

The Ministry of Health, in 1953, set up the Institute for Postgraduate Medical and Pharmaceutical Education in Prague, and in 1957, an independent institute was inaugurated at Bratislava. The main teaching units of the institutes are chairs and or departments covering the basic specialities which may later be divided into sub-departments for the more important of the highly specialized disciplines. The two institutes do not have their own hospitals, but use research institutes and the special departments of the larger Prague hospitals. The chair of clinical biochemistry is situated in the municipal hospital in Prague.

Basic specialist training
Basic postgraduate medical training consists of independent study while working in a special department under the supervision of its chief, an experienced specialist. To ensure a uniform standard of training throughout the country, the chairs of the institute have drawn up standard programs for all specialities and grades as well as model plans and teaching guides. Making use of these under the supervision of the chief of the department and in the light of the circumstances where he is working, each postgraduate student has to work out his own program of studies for the whole period of training in his speciality.

So that doctors may have an opportunity to develop and improve the knowledge and experience acquired in their department, the government has recently accorded them the right to paid leave for two to three weeks a year (for practical work at selected places, consultations and study prior to sitting for their examinations) if, in the same year, they have not taken part in a supplementary course organized by the Institute of Postgraduate Medical Education.

If a doctor has no opportunity to study his speciality in his department, he can study for a spell in another department, as a rule in the same region. Permission for this is given by the appropriate district and regional specialists. In the districts and regions group teaching facilities, including seminars, conferences and courses, can be organized for doctors taking specialist training.

Higher specialist training
Additional specialist training facilities at a higher level are provided by the Institutes of Postgraduate Medical and Pharmaceutical Education in the form of courses for groups of 15 to 25 students lasting from one to three months. These consist mainly of theoretical lectures. For individual teaching, groups consisting of one to four students are arranged with a mainly practical syllabus in an appropriate department for a period of one to three months.

The training of the specialist ends in a qualifying examination by a committee appointed by the institute. The committee consists of a member of the institute's staff, another outstanding specialist and a representative of the National Board of Health. The first-grade examinations in the basic specialities usually last one day, those for second-grade qualifications and higher specialities some two to four days. Candidates are asked questions on theory and practice, mainly in oral interviews, though for some specialities short written answers are required. After passing the examination, the candidate is awarded a specialist's diploma.

Further training and refresher courses
For the second aspect of postgraduate training and the further improvement of the specialist's qualifications, the Institutes of Postgraduate Medical and Pharmaceutical Education have also tried to work out their own approach, in keeping with the capacity of the Health Service and their own requirements.

Here, even more markedly than in the first type, emphasis is placed on in-service training under the supervision of a senior staff member and on the availability of district, regional and national study facilities. The leading role is played by regional and district specialists or their colleagues. The institutes are also concerned with university qualification obtained elsewhere than at medical faculties. In our case it concerns chemists with university qualifications.

In the light of their experience, the institutes have introduced several types of course for the further training of specialists, among which three regular training courses for medical teachers (i.e. the chiefs of all hospital specialist departments) are of outstanding importance in providing instruction on the latest advances in theory and practice. The institutes organize such courses every three years and they last two to three weeks. At lectures, seminars and demonstrations, the latest achievements and experience in the branches concerned and allied disciplines are explained and the participants then discuss questions of organization and teaching methods in their disciplines. Those attending these courses are responsible for sharing the knowledge acquired with their colleagues and those under them.

Another type of course organized often, and enjoying great popularity, is that on a special subject for senior and junior specialists interested in some particular new aspect of their discipline. Those invited to attend are mainly doctors who are in a position to introduce new methods and promote their adoption in their place of work.

The institutes annually issue a review of all their activities, indicating conditions of study and quotas for the guidance of regional specialists in recommending candidates. The final decision about acceptance of candidates is taken by the appropriate chairs at the institutes. All aspects of teaching are carried out free of charge. During instruction, each doctor receives his pay, his travel costs are met, free lodging is provided, and meals in the institutes dining room are available at cost price.

Throughout these postgraduate courses, leading specialists from the medical faculties, research institutes and hospitals in the locality, besides the staff from departments of the institutes take part in the lecture and in practical work, seminars, conferences, and working groups.

Place of clinical chemistry in non-medical education: technologists

Technologists and technicians working in clinical chemistry are trained in four-year schools. After three years practice they may achieve the specialist-biochemist grade. This education is provided in a special institute for medical technicians and nurses, under the jurisdiction of the Ministry of Health.

National associations and journals

The Purkinje Medical Association is the national organization and provides independent sections for all specialities including clinical chemistry. The Association of Clinical Chemists, a member of the IFCC, is a society combining physicians and chemists interested in clinical chemistry. Under the protection of this association various meetings and congresses are organized. The association is not concerned with special examinations or professional qualifications for the subject.

A great help for training and education of our clinical biochemists is also provided by the activities of scientific societies, namely:

(1) Czechoslovaque Biochemical Society of the Czechoslovaque Academy of sciences. In the activities of this society there are also regular postgraduate lectures, usually six during a year dealing with actual problems and progress in biochemistry. Practical courses are arranged, usually twice a year, covering modern laboratory techniques/enzymology, protein chemistry and immunochemistry. These activities are open to all clinical biochemists who are members of the Biochemical Society.

(2) The Society of Clinical Biochemistry, a member of the scientific societies organized in the Czechoslovaque Medical Society of J. E. Purkyné has in its program regular scientific meetings at which recent advances in clinical biochemistry are presented. In this society laboratory technicians can also be members. In the last year a branch commission of biochemists (non-medical) has been opened. The commission prepares refresher courses on recent advances in clinical biochemistry once a year.

Of great help for education and training is also the biochemical journal, *Biochemia Clinica Czechoslovacca,* edited by *Obzor* in Bratislava. The journal appears four times in a year.

There is a great disadvantage for teaching clinical biochemistry in that this important part of clinical education is not included in the teaching program of the medical curriculum of the faculty. There is general biochemistry in the preclinical part of education but no clinical biochemistry in higher terms where clinical subjects are taught. Information of significance of biochemical findings for diagnosis and differential diagnosis is notably absent.

DENMARK

History
Although minor hospital laboratories were established in Denmark around 1900, only a few real departments of clinical chemistry appeared before the Second World War. The growth began in the fifties, and 28 departments existed in 1960. In 1972 the number had increased to about 40. They are all headed by MDs. A further 48 hospital laboratories are supervised from these departments, while some minor laboratories are directed by heads of departments of internal medicine.

Legal status
The speciality 'clinical chemistry and laboratory technique' was established in 1946. In 1966 it was divided into the separate specialities 'clinical chemistry' and 'clinical physiology'. The present report only pertains to the speciality 'clinical chemistry'.

The legal status of clinical chemistry is based on the official status of the specialization in clinical chemistry of MDs with a stipulated training in clinical chemistry. They acquire the officially acknowledged status of 'specialist' in clinical chemistry by application to the Department of Public Health under the Ministry of the Interior.

Education and training for directors and senior staff of clinical chemistry laboratories

Medical graduates
At present, approximately 100 MDs have obtained a certificate in clinical chemistry in Denmark.

The requirements for obtaining a certificate in clinical chemistry are as follows:

One year's training in a department of clinical chemistry headed by a professor in the subject. The departments are at Rigshospitalet (Professor P. Astrup), University of Copenhagen and at Kommunehospitalet (Professor R. Keiding), University of Arhus, in the near future also: Odense Sygehus (Professor M. Hjelm), University of Odense, Amtssygehuset i Herlev (Professor O. Siggaard-Andersen) and Sygehuset i Hvidovre (Professor S. Müllertz), University of Copenhagen. Three years of full service as physician at departments of clinical chemistry, of these three years only one year may be at a smaller laboratory. At least one of the years must be at a hospital other than the respective university hospitals mentioned above. One year of full service at a clinical laboratory other than a department of clinical chemistry, or at a relevant university institute. One year of full service at a department of internal medicine. Participation in a course in clinical chemistry of about 250 hours taught over a period of some three years. This course started in 1972 so that specialization granted to MDs from 1975 encompass this systematic teaching.

Non-medical graduates
No systematic training system for the graduates in pharmaceutical chemistry, chemistry, natural sciences, etc. working in departments of clinical chemistry exists at present. However, some courses are available to graduates in pharmaceutical chemistry and a committee under the auspices of the Danish Society for Clinical Chemistry is elaborating on the subject.

Place of clinical chemistry in hospital service
At present, about 40 major hospitals (11 in the Copenhagen area) have created separate departments of clinical chemistry, headed by MDs, certified as specialists in this field.

Further there are two major and a few minor privately-owned clinical chemistry laboratories in Copenhagen.

The responsibilities of departments of clinical chemistry headed by a specialist in the subject comprises routine chemical analyses and hematology for ward patients in the hospital and outpatients treated by the clinical departments. Thus the speciality 'clinical chemistry' is separate from specialities covering clinical microbiology, serology, morbid anatomy, and clinical physiology. Nuclear medicine, however, is usually attached to the clinical chemistry laboratory although some functions may be attributed to the radiology department, or the department of clinical physiology.

At many clinical chemistry laboratories, however, the above-mentioned distinction is not complete, so that the clinical chemist also may be responsible for microbiology, serology, and clinical physiology. However, by far the major part of all microbiological and serological work is performed by The State Serum Institute and its affiliated departments in Arhus, Odense and Copenhagen.

All graduates in the department of clinical chemistry function to some extent as consultants to the clinical departments in their special fields of interest, e.g. toxicology, enzyme chemistry, etc. However, the major part of this responsibility remains with the specialists in clinical chemistry (MD) who may have weekly or monthly conferences with the staff of ward departments, especially departments of internal medicine.

Place of clinical chemistry in non-hospital service
A major part of the laboratories outside Copenhagen cover analytical service to the general practitioners of the area. In the Copenhagen area, a great deal of the general practitioners are covered by privately-owned laboratories.

Place of clinical chemistry in medical education
The departments affiliated to a university have teaching responsibilities for medical students. The curriculum comprises one week's stay for teaching in a department of clinical chemistry. In addition a number of lectures are given in the first and fourth semester of the clinical training.

Place of clinical chemistry in non-medical education
Teaching of technologists remains with departments of clinical chemistry authorized for education by the Ministry of Interior, and two Central Schools. Some less systematic teaching and training of technologists and graduates in the form of lectures, minor courses, etc., in special subjects is most often prompted by the application of new fields of interest.

Technologists, technicians and other laboratory workers
[types and training]
Since 1958 the hospital technicians have been trained according to regulations issued by the Public Health authorities. This leads to certification as a registered hospital laboratory technician.

The trainee must have had at least ten years of basic education followed by a diploma examination on high school level with a complete curriculum in mathematics, including basic arithmetic.

The education in clinical chemistry takes a total of three years including:

(1) A two-to-four months period of practical training in an authorized department of clinical chemistry.

(2) Three months theoretical course at a Central School for hospital laboratory technicians (about 300 hours).

(3) Practical training in an authorized department of clinical chemistry for a period of nine to twelve months.

(4) A course at a Central School for a period of about eight months (about 850 hours) ending with an examination.

(5) Final practical training in the authorized department of clinical chemistry finished by a practical examination.

This education program only pertains to the clinical chemistry technicians.

In 1968 the histology technicians, and in 1969 the blood bank technicians, obtained individual regulations, both worked out according to the regulations for the clinical chemistry technicians.

Chief technicians and instructors receive an additional education for a period of nine months after having worked as laboratory technicians for a minimum period of two years.

At present the Danish Association of Medical Laboratory Technologists counts a total of 3,500 members including about 400 students.

National societies, associations and journals

Organization of Danish Laboratory Physicians under the Danish Medical Association
Founded April 29th, 1948. President: Poul F. Moller, MD, Medicinsk Laboratorium A/S PO Box 2, DK-1304 Copenhagen K. Number of members: 75.

This organization takes care of the professional interests of its members.

Danish Society for Clinical Chemistry
Founded in October, 1956. President: Henrik Olesen, MD, Department of Clinical Chemistry, Bispebjerg Hospital, DK-2400 Copenhagen NV. Number of members: 365.

Concerned with the scientific and educational aspects of clinical chemistry.

Scandinavian Society for Clinical Chemistry and Clinical Physiology
President: Kjell Jacobsson, MD, Department of Clinical Chemistry, Lasarettet, Hälsingborg, Sweden. Encompasses the societies of clinical

chemistry in Sweden, Norway, Finland, Iceland, and Denmark. Co-ordination body for the national societies.

Organization of Pharmaceutical Chemists in Clinical Chemistry under the Danish Pharmaceutical Association
Founded July 1st, 1964. President: Jens Rasmussen, M Pharm Sc, Centrallaboratoriet, Skanderborg Sygehus, DK-8660 Skanderborg. Number of members: 70.
Takes care of the professional and economic interests of its members.

The Danish Association of Medical Laboratory Technologists
Founded 1948. President: Eva Munck, Teaching Technologist, Magstraede 5, DK-1204 Copenhagen K. Number of members: 3,500.
Takes care of the professional and economic interests of its members.

Scandinavian Journal of Clinical Chemistry
Scientific journal edited by Scandinavian Society for Clinical Chemistry and Clinical Physiology, appearing eight times annually. Managing Editors: L. Eldjarn, MD, Professor, and S. Kiil, MD, Professor, Editorial Office: Institute of Clinical Biochemistry, Rikshospitalet, Oslo 1, Norway.

Economic status of clinical chemists
The salary, pension and fringe benefits of clinical chemists functioning as heads of departments is fixed according to negotiated settlements between the state and county authorities and the Medical Association. These also apply to MDs functioning as heads of privately-owned laboratories. For specialists in clinical chemistry, who are not heads of departments, the salary follows the lines stipulated in settlements between the Association of Junior MDs and the state or county.

ECUADOR

History
Education in clinical chemistry was initiated in Ecuador in 1950 with the reorganization of the University School of Chemistry and Pharmacy into the separate Schools of Biochemistry and Pharmacy. Modern programs of study including one in clinical chemistry were implemented. The first group of doctors in biochemistry qualified to work in laboratories of clinical chemistry were graduated in 1955. At the same time a laboratory of clinical chemistry was organized in the chemistry faculty of the Central University in order to provide supporting laboratory services for students enrolled in the medical services of the university.

The activities of this laboratory were expanded in 1961 to offer services for patients of small hospitals which did not have clinical chemistry analyses performed entirely by physicians. As a group the physicians had not had formal training in clinical chemistry.

Legal status

In April, 1969, the decree 8840 promulgated by the President of the Republic authorized the Public Health Ministry to establish regulations for physicians and biochemists working in clinical laboratories. In addition, these individuals are regulated by the rules of their own professional associations. In hospitals it is required that the director be a physician who works co-operatively with the biochemists. Presently almost all chiefs of clinical laboratories are doctors of medicine. Only in small cities are biochemists functioning as directors of the laboratories. Despite the national law which requires biochemists on hospital laboratory staffs, the present restrictions of budget and manpower have caused delay in the implementation of this provision. The present regulations do not establish minimum requirements or provide for certification for either physicians or biochemists who perform clinical chemistry analysis.

Education and training for directors and senior staff of hospital clinical chemistry laboratories

At present there are no centers for the education and training of directors and senior staff of clinical chemistry laboratories. After working for some time in an appropriate laboratory environment one may be promoted to such a position. The chemistry faculty of the Central University (Quito) has now introduced a program for the training of biochemists for clinical biochemistry. Some physicians and biochemists have been able to qualify by attending approved courses at the postgraduate level in foreign countries.

Place of clinical chemistry in hospital services

All the hospitals in the big cities have clinical laboratories. These laboratories are under the direction of medical doctors. Some of them in the provinces are directed by biochemists. It would be an ideal situation to have biochemists in all of these hospitals as heads of chemistry sections.

Place of clinical chemistry in non-hospital services

In Ecuador there are many private medical laboratories for the performance of all kinds of clinical chemistry analyses. In the two biggest cities, Quito and Guayaquil, there are probably about 100 laboratories. Approximately 50% are owned by medical doctors and the other 50% are owned by biochemists. In the small cities, there are few laboratories and the majority belong to biochemists.

We may say that in Ecuador, 50% of the laboratories are owned by biochemists.

Place of clinical chemistry in medical education
Students of medicine take the subject of biochemistry during four months of the third year of their studies.

Place of clinical chemistry in non-medical education
Students of the school of biochemistry receive two years training in biochemistry and two years in clinical chemistry during six years of studies. In addition, they have the obligation to obtain practical experience for at least six months. They must also complete a thesis on any theme in clinical chemistry. Students of the sciences do not take part in clinical chemistry analyses in the hospitals.

Technologists, technicians and other laboratory workers
Three years ago, in two universities of Ecuador, courses for technicians were started. After two years of study, students are awarded a diploma which qualifies them to work in the hospital laboratories. Unfortunately these schools do not have qualified teachers.

Possible future developments
It is very important to improve teaching for biochemists by setting up courses at the postgraduate level. It is hoped that the IFCC will help to improve the education programs. It is also necessary to find new opportunities for biochemists to collaborate in all the hospitals.

National societies
The Ecuadorian Biochemistry Association was created by decree published in the Official Register No. 50 of November 11th, 1968. The organization was founded by biochemists of Quito to promote the professional and technical development of clinical chemistry in Ecuador. The association, governed by a four-member board, (president, vice-president, treasurer and secretary), is an independent organization. The board meets monthly and the assembly at least once each year. The association has been a member of the IFCC since 1969.

EGYPT

History
The development of clinical chemistry in Egypt had actually begun towards the end of 1930 in the department of clinical pathology Kasr-el-Aini faculty of medicine, Cairo. Clinical chemistry at that time was restricted to a limited member of simple analyses which were carried out in the routine hospital laboratories. In private practice, some laboratories were set up mainly in Cairo and Alexandria; a few were directed by medically-qualified persons, but the majority were under pharmaceutical chemists who were skilled in the various analytical processes.

Clinical chemistry was practiced in such laboratories together with hematology, serology and microbiology.

Tuition and qualifications

At present, in Egypt, there are eight universities with eight faculties of medicine, six faculties of pharmacy and eight faculties of science. Most of these faculties include departments of biochemistry in which brief courses of clinical chemistry are given to the medical and pharmacy students. The first postgraduate qualification in clinical chemistry proper, was initiated in Egypt in 1942 when a 12-months' diploma course in biochemical analyses was started at the school of pharmacy, faculty of medicine, Cairo University. Candidates for this diploma were university graduates either in medicine, pharmacy, veterinary science or general science, who had at least one year of training in one of the university hospitals or an equally recognized analytical laboratory. Another diploma (two academic years), which covered most topics of clinical chemistry, was the diploma of medical sciences. This was undertaken by the biochemistry department, faculty of medicine, science, pharmacy and veterinary medicine.

Similar courses were later introduced in other universities, e.g. Alexandria (1947) and Assiut (1962) in Upper Egypt. Such courses have largely contributed towards the establishment of a clinical chemistry discipline in this country, and the diploma was recognized by the public health authorities as an offical qualification of registration and practice of medical chemistry for both medical and non-medical personnel.

Apart from this, clinical chemistry is now taught at two postgraduate levels in most medical faculties in Egypt only to medical graduates as general and special courses.

The general course is given as subsidiary subjects to the diploma and the master's degree candidates in general medicine, tropical medicine, forensic medicine, and general surgery. The special course, on the other hand is given over a two-year period together with hematology, microbiology and parasitology to candidates for the master's degree in clinical and chemical pathology. Clinical chemistry has always been considered, in this country, as a sector of the clinical pathology department, which includes hematology and clinical microbiology as well.

The establishment of the first department of chemical pathology in 1962 at Cairo University, signaled the recognition of clinical chemistry as an independent entity for which a central laboratory was set up in the Manial University hospital at Cairo. Although similar central clinical chemistry laboratories were also then established in Alexandria, Assiut, Al-Azhar, Ain-Shams, Mansoura, Tanta and Zagazig University hospitals, formation of separate departments in chemical pathology or clinical chemistry has so far not been achieved outside Cairo University.

Clinical chemistry in hospital services

Apart from the eight university hospitals, the health care of about 40 million inhabitants is provided by 190 provincial hospitals, 12 district hospitals, 23 specialized hospitals, 299 rural health centers and 1,300 rural health units.

Several private hospitals and clinics are located in big cities. Clinical chemistry is developed in only a few of the provincial and district hospitals in addition to the public health regional laboratories situated in five big cities. With few exceptions, most of these laboratories suffer from serious shortages in staff, equipment and satisfactory administration. Besides the well-known difficulties of developing countries, current international political problems have hindered the country's developments in various ways of which clinical chemistry is no exception.

The uncertainty of supplies with the very limited funds together with the lack of space and personnel, all add to the difficulties of these laboratories. In spite of the increasing demand and the urgent problems, most methods of analysis are manually conducted and not properly controlled.

Future prospects

As indicated in the resolution of the First Regional Arab Congress on clinical chemistry held in Cairo, April 13th to 17th, 1974, much more attention should be given to clinical chemistry in this critical part of the world in order to be able to participate and take an active role in the health care of the community. This has to be achieved through supporting the university status and by developing new independent and well-equipped departments. Diagnostic laboratories should be provided with adequate staff and appropriate equipment.

The health centers and health units in the rural area need to be supplied with suitable laboratory services which are hitherto lacking.

The creation of the Central Quality Control and Reference Laboratory has been warmly approved by health ministers of many Arab countries, and it is hoped soon to take the practical steps towards their formation. This will help to solve many practical as well as technical problems of clinical chemistry in this area.

Last but not least, the problems of the laboratory technologists and assistants should be given due care and attention. Much dependence, however, is laid upon the betterment of the economic situation of this country during the years to come.

Legal status

In 1954, the Ministry of Public Health, issued an official charter to organize medical laboratory practice in the Arab Republic of Egypt. According to this law four official registers were set up for: medical biochemists (clinical biochemists), microbiologists, clinical pathologists and pathologists. Only individuals with specific university qualifications are allowed to be registered and licensed for the particular medical laboratory practice.

Societies

The Egyptian Society of Clinical Chemistry was formed in December, 1970 and was officially registered and recognized on April 7th, 1971. The society has so far held eleven successful scientific meetings and was elected to the membership of the IFCC in 1973. The society organized the First Regional Arab Congress on clinical chemistry, April 13th to 19th, 1974 which was attended by over 200 members from different Arab States as well as guests from Europe and other countries. The membership of the society has now attained 140 members.

Subsidiary branches of the society are now being formed in Damascus and in Khartoum. The Second Regional Arab Congress on clinical chemistry took place in Damascus late in 1976.

Medical technologists

In Egypt there are now four technical high schools for medical laboratory technologists at Alexandria, Assiut, Cairo and Mansoura. A fifth one has just started in Tanta.

These schools, under the control of the Ministry of Public Health, supply the ministry hospitals and health centers with laboratory and x-ray technologists. Neither the quantity, nor the quality of the graduates of these schools meet the present requirements. Representatives from the ESCC are now in negotiation with the public health authorities to revise the three-year syllabus of these schools and to provide facilities for higher training courses affiliated to the universities. The total number of these technologists serving in the different hospitals all over the country is 1,855 at the moment. Besides these technicians there are over 5,000 laboratory assistants with intermediate qualifications who are working in the rural health centers and units. These technicians need special attention to raise their standards up to their responsibilities.

ETHIOPIA

History

Clinical chemistry in Ethiopia is in the process of development. A senior technician is in charge of most hospital laboratories under supervision of a clinician, who seldom has special training in laboratory medicine. An Institut Pasteur d'Ethiopie was established in Addis Ababa in 1952. Its major field of work was bacteriology, serology and vaccine preparation, but it also included a section of chemistry for food and water analyses. This section was headed by a doctor of pharmacy, and it started analyses in clinical chemistry as a service to the hospitals in Addis Ababa.

The Pasteur Institute was taken over a few years ago by the Ministry of Health under the name Central Imperial Medical and Research Laboratory.

Education for Clinical Chemistry

A Nutrition Research Institute was established in Addis Ababa in 1962. The biochemical laboratory of this institute is headed by a medical doctor with clinical chemistry as a speciality. It is well equipped and to a limited extent performs clinical chemistry analyses for a few hospitals in Addis Ababa.

The first medical school in Ethiopia was established in Addis Ababa during the years 1964-1967. A clinical pathology unit was organized in the teaching hospital during 1967 for both teaching and provision of laboratory services to the teaching hospital. The department was headed by a professor of clinical biochemistry, but academic staff members were also employed in morbid anatomy, histopathology and bacteriology.

Education for academic staff in hospital laboratories

At the present time there are no Ethiopian doctors trained in the speciality of clinical chemistry. A few Ethiopians have obtained a PhD degree in biochemistry after studies abroad, but they are all employed in preclinical biochemistry teaching or in research institutions.

The clinical pathology unit of the medical faculty has initiated a program for postgraduate education in clinical pathology subjects. The total period of training is four years, part of which is given at the clinical pathology unit of the Haile Selassie I University, and part in a suitable university department abroad. The training comprises clinical chemistry, hematology, morbid anatomy, histopathology, and bacteriology, with one of the subjects mentioned as a major field.

Place of clinical chemistry in hospital service

Most hospitals still have limited laboratory facilities. Routine blood, urine and stool examinations and a limited number of quantitative chemical analyses are performed. Only the few institutions mentioned above have facilities for more complicated clinical chemistry. The University Hospital is for the time being the only hospital in Ethiopia with a doctor specializing in clinical chemistry.

Place of clinical chemistry in medical education

Ethiopian medical students take a course in general biochemistry during the first year in the medical faculty simultaneously with the studies of physiology and anatomy. The second year is mainly used for laboratory medicine, i.e. clinical chemistry, hematology, morbid anatomy, histopathology, bacteriology and parasitology. The course in clinical chemistry and hematology comprises 45 lecture hours and 30 three-hour practicals and tutorials. The course aims at giving the students an understanding of the biochemistry of disease and making them familiar with the chemical tests and laboratory methods used in clinical chemistry and hematology as an aid in diagnosis and for control of treatment. The students are taught the principles of the methods, sources of errors and how to interpret the results.

Technician training

Before 1957 no organized technician training was available in Ethiopia. Some hospitals had expatriate senior technicians who trained Ethiopian technicians on the job.

The Ministry of Health started courses for hospital laboratory technicians in 1957. The course leads to a diploma as junior laboratory technician after two years of training. A few one-year courses have also been organized by the Ministry of Health for junior laboratory technicians to upgrade them to technician rank. It has, however, been difficult to obtain competent teachers for these courses.

The university has organized three-year courses for technicians in basic sciences to meet the university needs, but co-operation between the Ministry of Health and the university in the training programs for hospital laboratory technicians has been difficult to co-ordinate.

Possible future developments

In the near future the medical faculty of the Haile Selassie I University should be able to give a complete postgraduate education in clinical pathology. Due to lack of manpower, however, very few hospitals can expect to obtain a specialist in clinical pathology subjects including clinical chemistry in the foreseeable future. Two Ethiopian doctors are now obtaining postgraduate education in clinical pathology abroad sponsored by the university, and a third is expected to join the clinical pathology unit of the medical faculty in the immediate future.

The training programs for hospital laboratory technicians should be improved, and a co-operation between the Ministry of Health and the university to this effect is desirable.

FEDERAL REPUBLIC OF GERMANY

History

The development of clinical chemistry began in the mid-19th century, parallel to the development of clinical diagnosis on the basis of natural science. The well-known publication by Liebig, *Die Tierchemie . . . in ihrer Anwendung auf Physiologie und Pathologie [Animal chemistry . . . and its application to physiology and pathology—1842]* strongly influenced clinicians' way of thinking and led to the setting up of clinical laboratories engaged in the clinical analysis of body fluids. In 1856, it was Virchow who established a separate department of chemistry at the Pathological Institute of the University of Berlin. In 1863, the first director of that department, Felix Hoppe-Seyler was appointed to the world's first chair of physiological chemistry at Tübingen University. The development of clinical laboratories was advanced by von Frerichs in particular. Paul Ehrlich was one of the first directors of a clinical laboratory and in his capacity as assistant to von Frerichs in Berlin, he developed the Diazo Test, called after him, as well as the aldehyde reaction.

In the early twentieth century, there were laboratories at nearly all major hospitals, headed by a number of well-known biochemists (e.g. Otto Warburg in Heidelberg; Hans Fischer, Siegfried Thannhauser, Otto Neubauer, and others in Munich).

It was the task of those laboratories to assist physicians in their diagnosis and in the supervision of medical treatment. In the eyes of the medical world, clinical chemistry became a so called ancillary discipline.

Clinical laboratories, however, were assigned no function in the training of medical students. Institutes of physical chemistry or departments at physiological institutes held this responsibility. Institutes of physiological chemistry also took charge of the scientific training of clinicians striving for an academic career.

After the First World War, courses of instruction in clinical chemistry and microscopy were introduced at the universities. Such courses were conducted by clinicians. To this day, most universities keep their clinical chemistry courses within the framework of their teaching program of internal medicine.

Before the First World War there was as yet no special training for laboratory staff. After a period of introduction, semi-skilled staff and nurses acted as helpers. The initiative was taken by the Berlin Lettehaus in the training of medical technologists (laboratory and x-ray assistants). More complex analyses were carried out by clinicians in addition to their regular work, but in the last 50 years there have been many instances in which it was necessary to put complicated procedures into the hands of experts working in the physiological chemistry institutes of universities.

Development of clinical chemistry as an independent scientific discipline
The first development of clinical chemistry as an independent scientific discipline after 1945 was determined by the following main factors:

(1)   The rapid development of advanced measuring techniques and the progress of biochemical analysis.

(2)   A major breakthrough in fundamental medical research; certain changes in metabolism were now recognized as causing illnesses and therefore became very important for diagnosis.

(3)   The physician's need for a scientific substantiation of diagnostic symptoms furthered the trend towards the determination of as many parameters as possible in physiological and pathophysiological processes.

Over the past 20 years, all clinical laboratories at universities and municipal hospitals have increased their staff and equipment considerably, with an ever-growing share in the total medical diagnostics falling to clinical biochemical analysis. This development has led to a steadily increasing differentiation between medical and laboratory functions. As the directorship of clinical chemistry laboratories could no longer remain a scientific sideline for physicians, the necessity for the training of clinical chemists arose. In 1955, the Gessellschaft für Biologische Chemie (Society for Biological Chemistry) set up a commission to work out guidelines for the training of clinical chemists. This commission certified as 'Klinischer Chemiker' (clinical chemist) those members of the Gesellschaft für Biologische Chemie, who—from their training and scientific work—could furnish proof of specific experience in the comprehensive field of clinical chemistry.

In 1964, those members of the Gesellschaft für Biologische Chemie who held a certificate as clinical chemist, founded the Deutsche Gesellschaft für Klinische Chemie (German society of clinical chemistry) as a separate scientific body. At the same time and parallel to this development, the German Medical Assembly, Deutscher Arztetag, decided to introduce, as a separate career, specialists in clinical pathology.

Legal status

Training and practice in the field of clinical chemistry are only in part defined by law.

As to the training of medical students, the approbatory regulations provide a course in clinical chemistry to be undertaken after the *Physikum* (the premedical examination). Thus, clinical chemistry has also become a discipline at university medical schools. For the post-graduate education of medical or chemical juniors in the subject of clinical chemistry, there is as yet no legal provision.

The position of a director at a clinical chemistry institute or a hospital laboratory is not subject to any qualification required by law.

Much of the work to be done in the area of clinical chemistry is carried out by specialists in clinical pathology, whose training covers bacteriology, serology, hematology, clinical chemistry and internal medicine. That training as well as professional practice is subject to the code of conduct for the medical profession.

It is the task of specialists in clinical pathology to carry out laboratory analyses on behalf of practitioners. The great majority of these specialists work independently at their own laboratories. At a number of hospitals, specialists in clinical pathology are in charge of clinical laboratories.

The professions of medical or chemical technologist are protected by law and defined in terms of training examination and professional practice. The law requires lectures, courses and examinations in clinical chemistry and hematology for the training of medical laboratory technologists.

Education and training for directors and senior staff of clinical chemistry laboratories

In view of the lack of legal regulations, the Deutsche Gesellschaft für Klinische Chemie specified guidelines for the training of clinical chemists.

Following a successful completion of the obligatory training courses an official certificate as a clinical chemist will be issued by the Deutsche Gesellschaft für Klinische Chemie, attesting the qualification of the office of a director at a clinical chemistry laboratory.

Guidelines for the training of clinical chemists

(1) The clinical chemist is required to have complete knowledge of customary clinical chemistry laboratory methods, to be thoroughly familiar with general chemical and biochemical practice, to be capable of independently developing new methods, and to work at problems of clinical biochemistry. He must be fully conversant with clinical medicine.

(2) The career of a clinical chemist is open to medical, biochemical and chemical postgraduates (state examination or diploma). The training takes five years, and is meant to widen the medical background towards biochemistry, and the chemical background towards medicine. All training steps must be documented by certificates issued by training-approved laboratories (i.e. their directors).

(a) Two years of training must be devoted to clinical chemistry diagnosis. This part of the training must be passed entirely either at a university institute of clinical chemistry or at a clinical chemistry laboratory under a fully-qualified directorship. Up to three years may be spent at university institutes of physiological chemistry or biochemistry.

(b) At the same time, applicants should also acquire a sound knowledge of hematological work.

(c) By producing the relevant certificates, medical applicants must furnish proof of their attendance at a practical course in chemistry (chemical laboratory sessions for future teachers or scientists), at a course in physics and chemistry, and at the mathematical courses required of chemists.

(d) An equivalent training within the framework of biochemical studies instead of chemical studies and premedical training is also recognized.

(e) By producing the relevant certificates, chemical applicants must furnish proof of their attendance at those courses which are prerequisite to the preliminary medical examination [*Physikum*].

(f) Scientific work in the field of clinical chemistry and clinical biochemistry must be proved by pertinent publications.

Place of clinical chemistry in hospital service

As laboratory organization varies from hospital to hospital, the place of clinical chemistry in the hospital service is not always the same.

University hospitals are usually equipped with clinical laboratories of their own to carry out chemical and biochemical analyses. Many of these laboratories are under the direction of physicians or head physicians who at the same time work in the regular hospital service. At present, there are ten central clinical chemistry laboratories at German universities.

At municipal and private hospitals, the situation is similar to that of the university hospitals. About 20 out of the hospitals with over 600 beds have clinical chemistry institutes under a qualified directorship. Many clinical laboratories at small hospitals (with less than 400 beds) are under the direction of medical technologists.

Place of clinical chemistry in non-hospital service
In West Germany, clinical chemistry analyses are carried out on a large scale outside hospitals by physicians (general practitioners and specialists for internal medicine). It is estimated that approximately 25,000 physicians have small laboratories of their own. There is also a large number of private institutes of laboratory medicine, usually led by specialists for laboratory medicine. It is only in exceptional cases that clinical chemists work outside hospital laboratories.

Place of clinical chemistry in medical education
As described above, clinical chemistry is among the obligatory subjects for medical students. The universities have established clinical chemistry and/or clinical biochemistry chairs within their faculties of medicine. At the moment, there are 12 chairs, 9 of which are linked with a central clinical chemistry laboratory (München, Berlin, Giesen, Hannover, Ulm, Aachen, Göttingen, Hamburg).

Place of clinical chemistry in non-medical education
Up to now, at West Germany's universities, there is only one course of studies that includes clinical chemistry as an optional (elective) subject, namely that of biochemical studies at natural science faculties. Within these faculties there are no chairs or institutes for clinical chemistry.

Technologists, technicians and other laboratory workers
Clinical chemistry laboratories employ staff with different types of professional training as follows.

Medical technologists [Medizinisch-Technischer Assistent, MTA]
Two years of training after passing the 'Mittelschule' (non-classical secondary education). At the end, there is a state examination. Some 90% of the MTAs are female. The training is equivalent to the 'B' level in the World Health Organization classification.

Chemical technologists [Chemisch-Technischer Assistent]
Three years of training, followed by an examination under state supervision.

Doctor's secretary
One year of training (or three years of apprenticeship), followed by an examination before the local medical authorities.

Laboratory assistant
Three years of training, followed by an examination before the Chamber of Industry and Commerce.

In addition, laboratories also employ semi-skilled helpers without any specific training. On an average, young women will work no longer than about two years. As on the one hand existing training centers for laboratory staff offer no adequate capacities and on the other hand, there is an ever-increasing number of analyses to be carried out, laboratories suffer under a shortage of skilled staff.

Possible development for the next ten years
The future development of clinical chemistry in the Federal Republic of Germany will be determined mainly by the progress in automation. The urgent need for rationalization leads to the building-up of central clinical laboratories. We may assume that within the next ten years, all university hospitals will have clinical chemistry laboratories doing their own research. How far municipal hospitals will establish clinical laboratories, depends on their size. Presumably, hospital authorities will consider central laboratories at hospitals with approximately 700 beds to be profitable.

Future development leads to a functional division between laboratories according to equipment. Complicated analyses such as those in the areas of endocrinology and enzymology will not be feasible nor profitable unless carried out by relatively large central laboratories.

The necessity of rationalization implies a standardization of laboratory methods and the application of computers. The Deutsche Gesellschaft für Klinische Chemie co-ordinates the activities of its members in various committees, which negotiate with ministries and governmental bodies for uniform development of laboratory equipment as well as training of relevant staff.

In the future, the development of clinical chemistry will particularly depend on the organization of clinical laboratories. The development of clinical chemistry as a separate scientific field is inseparably linked with the clear definition of what the clinical chemist's profession is and means.

National societies and journals

In the Federal Republic of Germany, clinical chemistry is represented by the Deutsche Gesellschaft für Klinische Chemie. Admission to this society is open to all certified clinical chemists or university graduates who can furnish proof of several years of practice in clinical chemistry. At present, the society has some 360 members. It is a member society of the IFCC and of the Deutsche Zentralausschus für Chemie (German Central Committee for Chemistry), and therefore also of IUPAC.

The Deutsche Gesellschaft für Klinische Chemie has the following objectives in mind:

(1)   In specific areas, co-operation with other medical societies and with the Gesellschaft für Biologische Chemie.

(2)   Promotion of training and education in clinical chemistry and clinical biochemistry at the university level.

(3)   Training and education of capable young scientists.

(4)   Co-operation of members in clinical and biochemical research and solving the problems of organization of clinical laboratories.

(5)   Creation of a recognized vocational status for clinical chemists, with a view to an appointment of qualified directors to all large clinical laboratories.

(6)   Conferences in all areas of medicine and biochemistry, in close co-operation with other scientific societies.

In addition to the practical work done by the Deutsche Gesellschaft für Klinische Chemie's executive board, there are a number of committees engaged in problems of standardization, training, medical technology, and in the planning of conferences. For its own members and together with other West European societies, the Deutsche Gesellschaft für Klinische Chemie organizes interlaboratory tests for quality control.

Since the Deutsche Gesellschaft für Klinische Chemie was founded, it has organized each of the following important conferences: the International Congress on Clinical Chemistry (1966) Munich; the West European Symposium on Clinical Chemistry (1967) Berlin; and the Congress on Biochemical Analysis in Munich, in co-operation with the Gesellschaft Deutscher Chemiker, the Gesellschaft für Biologische Chemie, and the Deutsche Pharmazeutische Gesellschaft (so far, in 1968, 1970, 1972, 1974, 1976). This is a regular conference taking place every two years, accompanied by a major exhibition of scientific equipment. In 1974 the Congress was held jointly with the First European Congress of Clinical Chemistry.

Subject to training regulations, the Deutsche Gesellschaft für Klinische Chemie grants the certificate as a clinical chemist. *The Journal of Clinical Chemistry and Clinical Biochemistry,* published by de Gruyter in Berlin, is the official organ of the Deutsche Gesellschaft für Klinische Chemie. It is published every month. In addition, the Deutsche Gesellschaft für Klinische Chemie publishes *Mitteilungen* every two months, these 'newsletters' are used to inform the society's members.

FINLAND

History
The first academic position in the hospital laboratories was created in 1930 at the Maria Hospital in Helsinki. The first laboratory physician with a clinical chemistry background held this position in 1941. Maria Hospital pioneered medical isotope investigations doing these in the later 1940s before radioisotopes were commercially available. Of early developments the Tallquist hemoglobin test may be mentioned as the first strip test. The first hospital biochemist entered hospital service in 1947.

The Finnish Society of Clinical Chemistry and Physiology was formed in 1947. It organized the Fourth Nordic Meeting in Aulanko in 1953 followed by others at eight year intervals. Teaching in clinical biochemistry commenced in 1959 in the science faculty of the University of Helsinki. The medical faculty created an associate professorship in clinical chemistry in 1965 and a full professorship in 1967. Presently there are five professorships (associates and full).

In 1971, the Finnish Society, together with other interested bodies, formed a company to promote the development of clinical chemistry and the performance of clinical programs and others on methodology and standardization of information.

Legal status
The conduct of clinical chemistry is governed by the Bill of Medical Diagnostic and Treatment Institutions (Laki Lääketieteellisistä tutkimus- ja hoitolai-toksista) issued in Helsinki 426/1964, statutes 442/1964, which governs the field of private institutions without beds, and private laboratories.

Laboratories in privately-owned institutions with beds are subject to the regulations of the Bill of Community Hospitals (Laki kunnallisista sairaaloista 561/1965) which, however, does not contain any specific statutes on the conduct of laboratories.

In paragraph 2 of the bill 426/1964 it is stated ' . . . medical examinations and treatment should be performed under the supervision of physician in charge approved by the State Medical Board (SMB)'. Pertinent points in the statutes 442/1964 specify that:

They apply to institutions that, without beds for patients, conduct for medical purposes:

(a)   chemical, serological, microbiological, histological, cytological or corresponding investigations;

(b) investigations or treatment using radioactive isotopes;

(c) other examinations for the assessment of state of health, diagnosis of disease, or prescription of treatment.

The physician-in-charge should be a legalized physician with the rights of a specialist in the appropriate medical speciality. The SMB may, however, approve as a physician-in-charge, a legalized physician without the rights of a specialist in the appropriate field, but who is regarded as being sufficiently familiar with the area; in which case the SMB should restrict, when issuing the licence, the examinations to be carried out and the treatment to be given in the institution, according to the qualifications of the physician-in-charge; in the case of chemical examinations the possible appointment of a clinical chemist should be taken into account.

In the institution there should be a sufficient number of personnel with professional education. The SMB may, when finding it pertinent, decree the minimal number of persons as well as their education that the institution should have.

The laboratory and corresponding apparatus and equipment in the institution should be calibrated before they are used, and they should be checked at intervals during use. The SMB may, if it finds it appropriate give more detailed decrees on this matter.

Records should be kept in the institution according to the detailed instructions issued by the SMB not only of the calibration of apparatus and equipment, and the checking thereof, but also of the laboratory and corresponding methods used, and of the control of analytical results.

To check methods and laboratory performance it is required that SMB samples be analyzed and results reported.

The legal status of specialities of medicine is defined in the Bill of Medical Practice 20/1925, as amended 308/1960. The bill restricts the right to use the name of 'specialist'. It states that the SMB issues the certificates for this, and issues regulations and specifications regarding the number and kind of specialities, how they can be obtained and the privileges attached to them. In this report the speciality 'clinical chemistry' will be dealt with in the section of education. It should be noted that clinical chemistry in the law text is regarded as a speciality of medicine.

Training and education

Medical graduates

The SMB has, through the Speciality Consulting Board (SCB), issued detailed instructions regarding training and education of physicians to qualify as specialists in clinical chemistry. These will be elaborated below. The regulations are in principle the same in the other Scandinavian countries. The training for a speciality involves:

(1) General training;
(2) Special training: (a) special training in the appropriate field;
(b) special training in related fields;
(3) Examination;
(4) General training is divided into: (a) hospital service (1 year);
(b) service in health centers (1 year).

For 'hospital service' ten months may be done as an intern, of which at least four months should be in internal medicine and two in surgery. These should be done after completion of all clinical courses. Six months is obligatory for service in health centers but six months may be exchanged for work as a medical officer in the armed forces or companies with more than 500 appointees. Optional training may be obtained in teaching and specialist training hospitals, in hospitals or departments headed by a specialist, except in the speciality for which the trainee prepares; in full-time posts in medical offices of the state in municipalities, in university medical theoretical departments, in medical state institutes, and as a full-time research fellow in a field approved by the SCB.

'Special training' lasts four years, of which three to three-and-a-half is in the appropriate field. The training is performed after legalization as a physician. The training has to be performed in teaching or specialist-training hospitals. In clinical chemistry the training should be performed in a clinical chemistry department of a hospital. However, service in university departments of medical chemistry may substitute maximally two years, while service in university departments of pharmacology, physiology, biochemistry, medical microbiology or radiochemistry, as well as a degree of candidate of philosophy, mastership in chemistry, biochemistry, physics or radiochemistry, may substitute for one year. Special training in related fields, a half to one year, may be obtained in a department of clinical physiology, clinical chemistry, clinical pharmacology, clinical microbiology, pathology, or in other laboratories approved by the SCB.

The 'specialist examination' is supervized by persons proposed by the medical faculties and approved by the SCB. An applicant who has failed twice in the examination, has to wait at least one year before he may take the examination again.

In clinical chemistry the requirements are as follows:
(1) Information contained in the following handbooks: White, Handler, Smith: *Principles in Biochemistry;* Henry: *Clinical Chemistry;* Stanbury Wyngaarden, Fredrickson: *The Metabolic Basis of Inherited Disease.*
(2) Knowledge of papers on clinical chemistry which have appeared during the last three years, in the following journals: *Duodecim, Finska Lakaresallskapets Handlingar, Nordisk Medicin, Suomen Laakarilehti, Advances in Clinical Chemistry, Clinica Chimica Acta, Clinical*

Chemistry, The Journal of Clinical Endocrinology and Metabolism, The Journal of Laboratory and Clinical Medicine, Metabolism and *The Scandinavian Journal of Clinical Laboratory Investigation.*

(3)   Knowledge on the basis of optional monographs of endocrinology, enzymology, hematology, isotope techniques, metabolism of water and electrolytes, and the basis of statistics.

Hospitals, in which specialist training can be performed are divided into 'teaching' and 'specialist-training hospitals'. Teaching hospitals are university hospitals. A specialist-training hospital may, in special cases and only for a specified term, be regarded as a teaching hospital, if its level corresponds to what is required for a teaching hospital. Hospitals in which four specialities are represented, may be approved as specialist-training hospitals, as well as specialized hospitals, departments or institutions. The teacher should be a scientifically-qualified experienced specialist in the appropriate speciality. The patient population and analysis material should be sufficient and suitable for teaching purposes and the teaching should be organized in a way approved by the SCB, which may also approve service as a physician in such a hospital as part of the training for a speciality to the extent it finds appropriate.

Corresponding training in the Scandinavian countries is regarded equal to the Finnish training. Service in other countries is approved when found appropriate.

Exceptions from these rules can be made when the medical faculty has acknowledged the competence for a professorship (docent) in the appropriate field.

These regulations have been approved by the SCB on 18th April, 1966 and apply for physicians who were legalized after 1st July, 1966. For those legalized prior to this date, the regulations approved earlier were applicable until 31st December, 1974. These regulations will not be further elaborated upon here.

For specialist-training purposes the teaching and specialist-training hospitals have positions for residents, who are appointed for a term of three years.

For the teaching of clinical chemistry for medical students the Universities of Helsinki and Oulu have founded chairs in clinical chemistry and the Universities of Helsinki and Turku have associate professorships in clinical chemistry. These professors are also responsible for the specialist training in clinical chemistry at the University Central Hospitals in the three cities.

Training of clinical chemists without medical degrees

In this report the term clinical chemist denotes academically-trained chemists without medical degrees. In the Scandinavian and Finnish languages it has become colloquial to use the corresponding term for medically-trained clinical chemists. In Finland, the only Scandinavian country with a sizeable number of non-medically trained chemists in the field of clinical chemistry, a term literally translated as 'hospital chemist' is colloquial.

Postgraduate training corresponds to the specialist-training of physicians in clinical chemistry with the following differences:

(1) The guiding and examining body, the Clinical Chemistry Competence Board, is not named by governmental agencies as is the case for the Speciality Consulting Board, but by the Association of Finnish Chemists (AFC), except for one member named by the SMB. Two members are regarded as representing the AFC, one of whom is the president of the section of clinical chemists, while two members are chosen from the university teachers.

(2) The training in the department of clinical chemistry at a hospital is three years. In the case of biochemists the required time is two-and-a-half years; if graduated from the clinical or clinical-analytical biochemistry line it is two years. One year can be substituted for by service in a department of biochemistry or corresponding institute; however, a minimum of two years in the clinical chemistry department of a hospital is required. Presently, 15 training posts as assistant clinical chemist appointed for three years are available, of these, 11 are in university hospitals.

(3) The literature recommendations for the examination are as follows: Davidsohn and Henry: *Clinical Diagnosis by Laboratory Methods,* W. B. Saunders Co., Philadelphia, London, Toronto, 1974. Curtius and Roth: *Clinical Biochemistry, Principles and Methods, Vol. I and II,* Walter De Gruyter, Berlin, New York, 1974.

Acquaintance with articles which have appeared during the last three years in *Adv.Clin.Chem., Am.J.Clin.Pathol., Clin.Biochem., Clin.Chim.Acta, Clin.Chem., Duodecim, Scand.J.Clin.Lab.Invest., Standard Meth. Clin.Chem., Z.klin.Chem.Biochem.*

Additional information from monographs on enzymology, acid-base and electrolyte balance, endocrinology, immunology, drug analysis and statistical methods in clinical chemistry.

These regulations became effective in 1968 and have been slightly amended a number of times. A file of certified clinical chemists is kept by the SMB.

Place of clinical chemistry in hospital service
The central laboratory is divided into different departments (clinical chemistry, hematology, microbiology plus serology and in some hospitals nuclear medicine). Often the physician is a head physician of a status equivalent to the clinical chief of physicians. This gives the laboratory and thus clinical chemistry an independent position in the hospital. In smaller hospitals there is often no laboratory physician. The chief physician of the hospital or the chief physician of internal medicine is legally in charge of the laboratory, though the hospital chemist, in practice, acts as the director.

The central laboratories usually do not perform histological and cytological examinations, which are done in laboratories directed by specialists in morbid pathology. Clinical physiology is still poorly developed, and may in part be included in the program of the central laboratories.

The largest university hospital laboratories, those of Helsinki, Oulu and Turku, have a large staff of academically-trained people.

Place of clinical chemistry in non-hospital service
Clinical chemistry examinations and analyses performed by private laboratories supplement the service given by the central laboratories. There are only a few private laboratories on a high professional level with a diversified program and a high technological capacity.

Place of clinical chemistry in medical education
Inasmuch as clinical chemistry in Finland is regarded as a speciality of medicine by law, training and education in the field is provided within the medical faculties. The programs for specialization have been described previously. As a preliminary to specialization and as a regular component of medical education, various facets of clinical chemistry are presented during the undergraduate years of medical education.

Place of clinical chemistry in non-medical education
In Finland, as described previously, it is possible for non-medical graduates in science to proceed with continued university and hospital programs to qualify as senior hospital chemists under the guidance and approval of the Clinical Chemistry Competence Board. As a preliminary to such advanced education and training an undergraduate curriculum in science may be followed. The faculties of natural science of the Universities of Helsinki, Turku, and Oulu teach clinical biochemistry or the equivalent clinical-analytical biochemistry, as one optional branch of biochemistry.

However, recently a curriculum specially planned for a career in the health service leading to a mastership in clinical biochemistry has been available in the newly-formed University of Kuopio. This curriculum is partly identical with the medical curriculum. During the first year there are common courses in basic chemistry and physics, cell biology and human biology and pathophysiology. The second year is mainly devoted to chemistry, but there are also courses on information service and laboratory data analysis. The third year is devoted to radiochemistry, analytical chemistry, protein and enzyme chemistry. During the fourth year there are courses in biochemical control mechanisms and the students carry out their MSc thesis work and pass their final examinations. There are in addition optional courses and the students are encouraged to practice in clinical chemistry laboratories.

Textbooks in biochemistry for the clinical-analytical line
At the 'cum laude' level (two years of study):
West: *Textbook of Biophysical Chemistry.*
Karlson: *Textbook of Biochemistry.*
Thimann: *The Life of Bacteria.*
Dawes: *Quantitative Problems in Biochemistry.*
  At the 'laudatur' level (final examinations):
White, Handler & Smith: *Principles of Biochemistry.*
Martin: *Introduction to Biophysical Chemistry.*
Gutfreund: *An Introduction to the Study of Enzymes.*
Hoffman: *The Biochemistry of Clinical Medicine.*
Hess: *Enzyme in Blutplasma.*

Technologists, technicians and other laboratory workers
Education of technologists comprises two-and-a-half years. The representative curriculum is from the Helsinki School of Medical Technologists. There are some differences between different schools, but the SMB exerts control by issuing licences to graduated technologists and keeping a file of them. The positions held are called 'laboratory nurses'.

By attending additional courses, including administrative topics, technologists may qualify for higher positions.

The curriculum in the School of Medical Technologists in Helsinki
(1) Medicine and therapy (general nursing, general pathology, internal medicine, contagious diseases, pediatrics, surgery, psychiatry, pharmacology)—210 hours.

(2) Mathematics and science (mathematics, physics, chemistry, statistics, automatic data processing)—400 hours.

(3) Sociology and psychology (and social policy, communication, English, education, administration, legal aspects)—170 hours.

(4) Laboratory sciences (biochemistry, clinical chemistry, clinical physiology, clinical microbiology, hematology, nuclear medicine, instrumentation, serology, hospital hygiene)—1,150 hours.

(5) Health service (public health service, ergonomy)—100 hours.

Formal theoretical teaching comprises 1,000 hours, laboratory exercises and practice 2,200 hours, altogether 3,200 hours.

Possible future developments
Number of academically-trained people
Presently about 60 clinical chemists and 84 laboratory physicians are active in the field, of these 41 chemists have been certified in clinical chemistry. The laboratory physicians are divided among the following specialities: pathology 25, microbiology 15, clinical chemistry 15, clinical physiology 9, clinical pharmacology 8, clinical hematology 4 and with no particular field 8. A survey carried out in 1972 on the creation of new positions in already existing clinical chemistry and related laboratories

revealed a need for about 30 physicians and an equal number of clinical chemists in 51 hospitals. Probably these positions will not all be granted immediately but the estimate may be met within ten years. An equal number of positions might be created in hospitals and laboratories to be built. That would bring the total number up from 80 to approximately 200.

A professorship in clinical chemistry (MD) will soon be founded at the University of Oulu and, in addition, an associate professorship in clinical chemistry (MD) at the University of Turku.

Within ten years it is hoped that a department of clinical chemistry at the University of Helsinki can be built. At present the department is situated in the central laboratory of the Helsinki University Central Hospital, which has little space for teaching and research.

Specialization
This will inevitably increase with a creation of subspecialities. Some positions for chemists have recently been created in serological and immunochemical laboratories.

Hormone chemistry has also become a specialized field. The specialization pressure for laboratory physicians is still greater, resulting in increased numbers entering clinical laboratories.

Impact of mechanization, automation and data processing
During the last years the first hospitals have reached operational data processing techniques, and three others are entering this stage during the next three years. There are no automated laboratories, only a number of partially-mechanized ones. (Technicon Auto Analyzer). There are no definite plans for automation on a level which will make co-operation on a regional level possible.

Health surveys
A survey based on truck-carried x-ray, medical examination and central analysis of a number of serum and urine constituents has been operational for three years. The limited number of cars and the low capacity of the Technicon equipment used made this effort insignificant on a national scale. Only a few small counties have been covered. The advent of more sophisticated techniques will probably speed up this effort.

Quality control
A laboratory control system encompasses all laboratories in Finland.

National societies and journals

The Society of Clinical Chemistry and Physiology in Finland
This is the scientific forum for clinical chemists of all categories. The society forms the link with international bodies.

The Section of Clinical Chemists of the Finnish Association of Chemists and the *Section of Laboratory Physicians of the Finnish Medical Association*
These bodies handle questions of professional and economic interest for its members.

The regional journal is the *Scandinavian Journal of Clinical and Laboratory Investigation,* sponsored by the Scandinavian Society of Clinical Chemistry and Physiology.

FRANCE

History
For fifty years due to Nicloux and Grimbert, and more recently, thanks to Polonovski, Fleury and Courtois, clinical chemistry in France has always been forward looking.

The numerous collaborators of these pioneers have created a French clinical chemistry area with a very broad basis; there are over 2,500 private laboratories in addition to the clinical chemistry laboratories of the public hospitals. While the latter specialize nowadays in research on particular problems of clinical chemistry or in other specialized fields, the former are distributed over the whole French territory and carry out a multivalent biological task, which includes at least 50% of clinical chemistry besides hematology, serology, bacteriology and parasitology. For this reason, as will be described in more detail later, the present training of clinical chemists provides individuals with the broad basic knowledge required to carry out routine examinations with maximum reliability coupled with an ability to adjust quickly to new technology. In addition the special needs of public hospitals have brought about the creation of teams of biochemists with a thorough knowledge of clinical chemistry who are able to follow rapidly the progress of this science while engaged in research programs in this field.

Legal status
The French 'Code of Public Health' defines the rules for the setting up of a private medical analysis laboratory. Book VII, under the heading no. III, which includes the law of the 18th May, 1946 defines the legal status of private medical analysis laboratories. The decree of 22nd February, 1965 relates to ante-natal and post-natal examinations concerning TB, syphilis, nephritis, heart disease, diabetes and feto-maternal antibodies.

The examination of blood incompatibility is carried out only in laboratories where the director holds the special certificate in hematology. Special provisions to carry out serology of syphilis are defined by the decree of 19th March, 1940. The Special Certificate of General and

Applied Immunology is mandatory. It is delivered by the faculties of medicine and pharmacy (decree of 20th July, 1965 which abolishes the former Certificate of Serology applied to Diagnosis of Venereal Disease, established by the decree of 16th November, 1948). According to the decree of 18th May, 1940, pregnancy tests can only be carried out by an officially approved laboratory. Finally, a decree of 16th June, 1967 gives a list of the minimum equipment required for all private medical analysis laboratories.

Although these regulations still are in effect, a proposal has been advanced by biologists together with the officials of the Ministry of Public Health to draft a new law concerning laboratories. This would provide that:

(1)   The office of director of a private medical analysis laboratory exclude any other professional job relating to medicine or pharmacy, thus officially creating the profession of medilab biologist.

(2)   The director of a laboratory should exercise this function personally.

(3)   He should be a doctor of medicine, a pharmacist or a veterinarian, and have at least three specialized certificates chosen from a list which will be published by decree:

> Bacteriology
> Clinical chemistry
> Hematology
> Immunology
> Parasitology
> Pathological anatomy (for medical doctors only).

Education and training for directors and senior staff of hospital clinical chemistry laboratories

For the time being the training of directors and senior staff of clinical chemistry laboratories in most cases is limited to the diploma of medicine or pharmacy with the special certificate of applied immunology and sometimes the certificates of hematology.

There is no obligatory specialized training for directors of clinical chemistry laboratories as most of them learn their speciality by practising in biochemistry laboratories and by obtaining different existing certificates in most of the large French universities.

Place of clinical chemistry in hospital services

The hospital services are classified in two categories: those pertaining to the faculty of medicine in the form of university hospital centers (CHU) and hospitals of smaller towns called secondary hospitals. In the university hospitals, the directors of laboratories or central laboratories who are head of departments must be doctors of medicine or pharmacists. They are generally professors of biochemistry in the faculty of medicine or pharmacy. They are nominated after competitive examination. This procedure has been started recently and does not function yet for the pharmacist. However, decrees which appeared in 1971 give us hope that medical biologists, doctors of medicine or pharmacists will be officially appointed to half of the services of clinical chemistry and professors of the faculty of pharmacy to the other half.

Place of clinical chemistry in non-hospital services
In France there are many private medical laboratories carrying out clinical chemistry analysis, hematology and microbiology. There are some 4,000 laboratories, small ones in secondary towns and others having 10-15 employees. All private hospitals with more than 200 beds have their own laboratories. The directors of these laboratories are gathered in several trade unions (syndicates), which are very active, and 1,500 medical biologists are members of societies of clinical biology. One thousand of these laboratories are located in large towns and some of them have up-to-date equipment. The number of laboratories associated with a pharmacy or a general practitioner will become very low, because the law will prohibit any new establishment of this type. Pharmacists having a pharmacy can carry out a few analyses, e.g. blood urea, blood glucose, and urine analysis. If they want to open a private medical laboratory, the law of the 18th March, 1946, obliges them to have special premises and equipment.

Finally, some highly specialized private laboratories have been set up in large cities, with modern equipment and qualified specialists. They collaborate to carry out accurate analysis for the whole town, at reduced expenses. In the future, the authorities should give consideration to the special problems of these laboratories.

A recent questionnaire showed that 70% of French private laboratories are directed by pharmacists.

Place of clinical chemistry in medical education
French medical students are taught general biochemistry during the first and second year of their study followed by a course on pathological biochemistry during the second part of their medical training.

Additionally, the decree of the 8th December, 1966, has established in France a set of studies and research in human biology.

The purpose of this set of studies is to train the senior staff of medical laboratories belonging to hospitals and the research teams who will contribute to the advancement in the different fields of human biology.

The decrees of 1967 and 1968 set up the training of physicians as medical biologists in three stages during their undergraduate medical studies, enabling the student to acquire a sound knowledge in human biology. Medical students are chosen after a preliminary examination, then during the six years of their medical study, they must take three specialized courses in a certain field and obtain the title of master in human biology. For example, in biochemistry the following courses with examination are demanded:

> General biochemistry
> Applied biochemistry or biochemical physico-chemistry
> Applied biochemistry relating to specific clinical fields,
>   (immunochemistry, endocrinology, genetics, etc.)

The third phase of the studies is essentially made up of formal teaching and of personal research. It is concluded by a thesis in human biology.

Place of clinical chemistry in non-medical education

Applicants belonging to the faculty of pharmacy or science (biological sciences) can acquire a master's degree and a doctorate in human biology and join the research or teaching staff.

These new rules have been followed since 1969 and it is too early to predict the impact on the training of clinical chemists. These provisions should insure the training of future heads of departments in hospitals. However, few students of the faculty of sciences are choosing this path. In fact the profession of director of clinical chemistry laboratory is, in France, practised only by physicians and pharmacists.

The draft (1967) of the law relative to private laboratories stresses the importance of postgraduate specialized examination, especially in clinical biochemistry. A special board has been set up by the Minister of Education, made up of physicians and pharmacists-biologists, in order to modernize the existing certificate. This certificate, open to medical doctors or pharmacists, comprises 150 hours of theoretical lectures, practical work, and a probation period in a hospital laboratory. Teaching is divided into three parts:

(1)   Principles and methods for physicochemical analysis in biochemistry.

(2)   Biochemical methods and their pathophysiological interpretations.

(3)   Biochemical basis of physiopathology.

Nearly 150 topical problems are studied. In the future, this certificate will be indispensable to open or direct a laboratory of clinical chemistry.

Taking into consideration the level of pharmaceutical studies in France (five years of university teaching after successful completion of secondary education) it is desirable that teaching of clinical chemistry will be carried out by the faculties of pharmacy and medicine. The program is the same in the two faculties. In Paris, because of the great number of students, the faculty of medicine and the faculty of pharmacy give their lectures independently. In the Universities of Lille, Strasbourg, Montpellier, Lyon, Toulouse and Bordeaux, the teaching for the certificate of biochemistry is carried out jointly for the student of both pharmacy and medicine.

97

Technologists, technicians and other laboratory workers

Technologists in France are trained in technical schools (non-university level) where young university assistants provide teaching in the different fields of biology. The training of technologists is an acute problem due to the low number of technical schools and the limited number of places available. Moreover teaching is very theoretical and not adapted to the developments in clinical chemistry. Teaching concerning electronic data control, automation, and quality control is usually missing in biological analysis (decree of the 5th August, 1954 of the National Education Ministry). This certificate is given after an examination organized by a board of examiners, consisting of qualified clinical chemists, university professors, professors of technical institutes and heads of hospital laboratories in Paris.

Possible future developments

During the last ten years considerable unplanned developments in clinical chemistry have attracted the attention of the government for several reasons:

(1)   Clinical chemistry is very interesting to numerous students in medicine, pharmacy or sciences, especially nowadays with too many university graduates.

(2)   Clinical chemistry costs too much when it is carried out by unqualified chemists. The government has to reduce the number of laboratories in order to insure highly-qualified directors and staff.

(3)   The quality of the analysis done and the reasons for carrying them out should be very exactly defined.

Consequently, we will see in the coming years efforts to organize the profession of medical biologists. In the future the medical biologist will have to prove his ability and efficiency in public hospitals as well as in private laboratories.

The draft of the new bill clearly defines and standardizes the training of clinical chemists. Their quality will depend on the standard of teaching.

National societies, associations and journals

Apart from syndicates which are busy with economic problems concerning medical biologists (either in private practice or hospitals) there are several scientific societies grouping medical biologists. All of them include medical biologists in different fields, and none are for chemists only. The French Society of Clinical Biology has a membership of 800. It was founded in 1942 by a group of biologists. The first chairman of the society was Professor Michel Polonovski. The society is ruled by a board of eight members assisted by twelve advisors, and is not affiliated to any of the other national scientific societies. Bi-monthly, a meeting is organized where the members may present their works (original manuscripts, laboratory techniques, general survey or a topic). Once a year, the society organizes a meeting devoted to the study of a specific biological topic.

The Society of Clinical Biology organizes symposia on clinical chemistry, conferences and demonstrations. The society is a member of the International Federation of Clinical Chemistry.

A journal is published bi-monthly by the society: *Les Annales de Biologie Clinique* edited by L'Expansion Scientifique Francaise, 15, rue Saint-Benoit, Paris 6°. The French Society of Medical Biology includes clinicians and medical biologists. Among French scientific journals should be mentioned, apart from the *Annales de Biologie Clinique, Pathologie et Biologie, European Journal of clinical and biological studies.*

GERMAN DEMOCRATIC REPUBLIC

History
In 1960, a Committee on Clinical Chemistry and Laboratory Diagnosis was constituted in the German Democratic Republic. This was the basis of the Society of Clinical Chemistry and Laboratory Diagnosis of the GDR, which was founded in 1967.

The Society of Clinical Chemistry and Laboratory Diagnosis—[GDR]
This society is a co-operative member of the Society of Experimental Medicine of the GDR which incorporates all experimentally orientated medical societies of the GDR.

Structure and aims of the society
The supreme organ of the society is the General Assembly of its members. The General Assembly decides all principal questions and elects the staff. The staff consists of the chairman and seven to ten members. The chairman of the society is at present Professor Dr. Haschen, director of the Institute of Clinical Biochemistry. Halle-Wittenberg, Halle (Saale). The election of the chairman takes place every second year. At present, the society consists of 395 members (227 chemists, 4 biologists, 121 physicians, and 43 technicians) who are working in hospital laboratories or in research institutions. Technicians who have been working for many years in the field of clinical chemistry are proposed by at least two ordinary members of the society and may apply for membership.

There are five subgroups of the society, organized on a regional basis (Berlin, Dresden, Erfurt, Halle, Rostock). The main task of these subgroups is postgraduate training in clinical chemistry. Furthermore, there are four special committees for the development of essential methodological problems (chromatography, electrophoretic techniques and diagnostic immunology, automation, electronic data processing and documentation). A further special committee (standardization and quality control) is institutionally attached to the 'Deutsches Institut für Arzneimittelwesen'.

As for research, members of the society co-operate in the following fields: diagnostic enzymology, other relevant problems of chemical pathology, automation and data processing, screening and presymptomatic diagnosis and the preservation of blood.

Congresses of the society are held every second year (the last one in 1975) in different towns of the GDR. The special committees organize symposia on a national or international basis. The regional subgroups gather three or four times a year.

Main tasks of the society

(1)   the advancement of scientific activity in research and practice and the exchange of experience in the field of clinical chemistry;

(2)   promotion of education in the field;

(3)   contribution to main problems of experimental medical research;

(4)   the support in transformation of scientific results into laboratory practice.

Standardization of diagnostic laboratory methods

In 1960 the Institut für Arzneimittelwesen, Berlin was charged with the task of standardization in clinical chemistry.

In the commission *Deutsches Arzneibuch* a committee on diagnostic laboratory methods was constituted. A specific volume containing diagnostic laboratory methods exists in the *Deutschen Arzneibuch,* 7th edition. The single methods may be supplemented any time by the addition of standard procedures for up-to-date methods. Standardized methods of the DAB (7th edition) are obligatory for all laboratories in the GDR.

The position of clinical chemistry in medical and non-medical education

The undergraduate education of the physicians in clinical chemistry occurs in several steps. The preclinical years of medical education provide for an introduction into the subjects in a course in biochemistry during the second year. Theoretical aspects of biochemistry are presented in lectures, seminars and periods of practical work in laboratories. The extent and considerable depth of this training in biochemistry is the theoretical basis of the second step of the clinical chemistry education. In the third year of education specific lectures and seminars in clinical chemistry and additional lectures are presented as a part of the course in pathology.

The postgraduate education in clinical chemistry extends for five years. The education includes a basic biochemical program performed in a research laboratory. Physicians acquire special techniques of clinical chemistry in modern clinical laboratories. This education is completed with an examination (specialist in clinical chemistry).

Chemists, biochemists and biologists are also engaged in clinical chemistry laboratories. Their specialization is provided in a five-year postgraduate program.

Medical technologists are educated in special technical schools in a three-year course. After at least two years of practical experience, training aiming at additional specialization (hematology, clinical chemistry, etc.), is possible.

GHANA

In Ghana 'clinical chemist' refers to a science graduate or medical doctor who has specialized in clinical chemistry; 'pathologist' refers to a medical doctor specializing in morbid anatomy; 'bacteriologist' refers to a medical doctor with specialized training in microbiology, and 'hematologist' refers to a medical doctor with specialized training in hematology and blood transfusion.

Legal status
A proposed Public Health Law will control the work of both private and public clinical laboratories.

Private clinical laboratories will be allowed under the direction of medical doctors or science graduates with recognized training in laboratory medicine, or medical laboratory technologists.

Societies
Professional societies or associations are recognized by law (NRCD 143).

There are the Ghana Medical Association, the Ghana Association of Medical Laboratory Technologists and the Biochemical Society of Ghana (yet to be accorded legal recognition).

School of medical laboratory technology
A technician's training school has been training medical laboratory technicians for the past 25 years.

It is proposed to upgrade this to a school of medical laboratory technology to train technologists as from September, 1975.

At the moment there are 45 medical laboratory technologists specializing in the following subjects:

Chemical pathology	7
Bacteriology	21
Hematology and blood transfusion	9
Histopathology	3
Parasitology	4
Virology	1

101

Health care in Ghana is provided at the Korle Bu Teaching Hospital, Accra, and central, regional and district hospitals as well as health centers and health posts throughout the country.

The teaching hospital, as well as the central and regional hospitals have the full complement of the relevant sections of a clinical laboratory. The teaching hospital and four of the regional hospitals are staffed by pathologists, medical officers, science graduates and medical laboratory technologists.

Except at the teaching hospital, the laboratories provide mainly service and little or no research is carried out.

The other regional and district hospitals laboratories are ill-housed, ill-equipped and poorly staffed.

The health centers and health posts have side room laboratories but few are manned by trained laboratory staff.

In addition to the government hospitals there are mission and private hospitals which supplement the health care services offered by the government.

Clinical chemistry in the university

There is a department of chemical pathology in the only medical school in the country. It has a teaching staff complement of one clinical chemist and two chemical pathologists. In addition there are three biochemists and six qualified technologists. The department has a unit which carries out normal routine service for the teaching hospital. It also serves as a reference laboratory for other laboratories in the country. The research unit of the department works in close collaboration with other departments of the school in its research programs.

Library facilities in the medical school are reasonably good and clinical chemistry is quite adequately catered for by way of the availability of journals dealing with the subject.

Medical education

There is at present only one medical school in the country, which has an annual intake of 75 made up of 65 medical and 10 dental students. The students undergo a course in basic biochemistry and pathology during the first three years. There is a formal course in chemical pathology during this period. The students gain some experience in routine laboratory work including routine urinalysis. There is a later program of integrated teaching which includes chemical pathology.

The medical course takes six years after secondary education. A second medical school at the University of Science and Technology is due to start admitting students from October, 1975.

Clinical chemistry—present and future

Clinical chemistry is still in its infancy in Ghana. Apart from the department of chemical pathology at the University of Ghana Medical School other regional hospitals only have clinical chemistry units as part of the department of pathology. It is proposed to start a course, in the near future, in clinical biochemistry leading to the degree of MSc. The course will be jointly run when started, by the department of chemical pathology and the department of biochemistry, both at the University of Ghana. As has been mentioned earlier, a technologist's school is being established to train technologists for the various laboratory disciplines including chemical pathology. The training of other technical staff is also being actively pursued. It is hoped that the program being followed will lead to a considerable improvement in the quality and scope of work of clinical chemistry laboratories in the country in the next ten years.

HUNGARY

History
Clinical chemistry has a considerable past in Hungary. At the end of the nineteenth century physicians in Hungarian clinics were conscious of the fact that modern medical science could break its century-old stagnation based solely on empiricism only by hand in hand development with chemical sciences and chemical research. The necessity for a professional organization dedicated exclusively to medical laboratory problems was felt soon after the Second World War. In 1946, a section for laboratory and experimental medicine, KOLAB, was founded by the Medical Division of the Trade Union. This section organized post-graduate courses in clinical pathology and meetings in which lectures on original work in the field could be delivered. In 1961, the Scientific Council of the Ministry of Health appointed a commission for clinical pathology. This commission elaborated (1961-1967) various topics of general interest concerning the organization and categorization of medical laboratories, the supply of instruments and chemicals, etc. A collection of 'recommended methods' has been prepared in 1964 and distributed to every laboratory in the country. In 1967, this commission finished work and its task has been transferred to the newly-founded Hungarian Society for Clinical Pathology which is a member of the Association of Medical Scientific Societies in Hungary.

The achievements of clinical chemistry in Hungary are known to the rest of the world through the work of many outstanding scientists. It is impossible to enumerate all of them, but the following names must be mentioned: G. Boroviczenyi, P. Hajdu, K. Laky, J. Selye, J. Somogyi, G. Szasz, A. Szentgyorgyi, A. Tarnoky, and Fr. Verzar. The work of the Koranyi School answered many questions about renal function and furnished a new basis for nephrology and therapy.

Legal status
The director of the clinical laboratories must be a physician as specified by the law of 1936.

Education and training for directors and senior staff of chemistry laboratories—physicians
Physicians having completed a three-year training course can pass a state examination before a board of examination, regulated by a law of 1936, on the condition that they spend at least two years in a medical laboratory. One year spent at a department for internal diseases or pediatrics can be taken into account. The examination is divided into practical and theoretical parts. The candidates have to give an account of their knowledge of physiology and general medical sciences. Besides their clinicochemical knowledge they have to verify a profound experience in bacteriology, serology and hematology. Upon completion of the examination the 'specialized laboratory physician' achieves a higher salary category and is authorized to apply for a leading post.

About two-thirds of all laboratory physicians are women.

The place of clinical chemistry in hospital and polyclinic services
The hospital laboratories are treated on equal terms with the other clinical departments. The chief of the laboratory department is always a head physician, subordinate only to the director of the institute. The staff of a laboratory consists generally of the following persons:

> Head physician
> Assistant physician
> Chemical engineers
> Pharmacists
> Head assistant
> Assistants
> Assistants without special schooling
> Administrator

About 95% of all technical personnel are women. The number and distribution of people of the staff depend on the number of beds, and on the type of the hospital. The theoretically determined number is generally nowhere attained because of the lack of qualified technologists and technicians.

The majority of the routine work consists of chemical analyses and hematological tests, but the bacteriological and serological examinations are for the most part within the scope of the laboratory duties. The histological examinations are generally the task of specialists in pathological anatomy. Joined to the clinical departments we find in many institutions the so-called 'department laboratories', which carry out the examinations not requiring a special qualification or instrumental equipment. So the central laboratories are released from urine analyses, blood counts, etc., which are carried out by semi-skilled nurses. In addition to laboratory work these nurses also have some nursing functions too. At the departments for internal disease, the blood pictures and urine sediments are examined generally by clinicians. From the operative departments these examinations are sent either to the internists, or to the central laboratory.

The equipment of the hospital laboratories is generally sufficient, whereas the laboratories in the polyclinics are not up to all expectations. The service laboratories work with home produced instruments. A preference is shown for the instruments of the Zeiss-Jena works, especially photometers, microscopes, and other optical instruments. The research centers, the laboratories of the university institutes and the largest hospitals can furnish themselves with the instruments of the well-known western factories.

The place of clinical chemistry in non-hospital or polyclinic services
There is no private laboratory practice in Hungary.

There is a tendency to organize so-called 'department laboratories' at the polyclinical centers, where urine analyses, some hematology and other simple determinations can be performed after the examination of the patients. The lack of sufficient qualified personnel causes some problems in the broader application of this program.

The place of clinical chemistry in medical education—undergraduate medical education
During the course of undergraduate medical education, clinical chemistry is presented in a non-integrated fashion with the subject matter spread over several faculties.

The Hungarian Health Ministry started to organize departments in the medical universities for teaching clinical laboratory diagnostics in undergraduate medical education. The first chair was established at the University of Pécs in 1975, directed by Professor Jobst. The second will be organized at the University of Budapest, presumably in 1977.

Postgraduate education
The program for the education of the physician laboratory specialists has been described in part previously.

In Budapest there is a High School for postgraduate education in medical sciences. About 15 years ago a department for clinical laboratory diagnostic was organized there. There, the physicians interested in laboratory diagnosis can spend some weeks to a maximum of three months, to perfect their knowledge in this field. Courses are organized in special themes (e.g. chromatography, hematology, immunology, etc.). Participation at these postgraduate courses is free of charge, but the physicians can take part only with the recommendation of their directors. They are given a paid holiday for this period.

The place of clinical chemistry in non-medical education
Two faculties of pharmacy (Budapest and Szeged) inaugurated a formal university course in clinical chemistry for undergraduates in 1975.

The final year students of the faculty have to select one subject of six options, clinical chemistry being one of the options. The course takes one year and the curriculum comprises lectures and laboratory sessions. At the end of the year there is a mandatory examination. Those who pass have a good chance to get a job in a clinical chemistry laboratory. It is planned that postgraduate training similar to that existing for medically-qualified individuals will be introduced.

Biologists, chemists and chemical engineers
There is no university training for clinical chemistry in faculties of science and chemical engineering. In spite of this several laboratories employ science graduates (mainly BSc, a few MSc) as 'biologists' or 'biochemists'. These graduates amass their knowledge during day by day work in the laboratory.

Assistants
In the Hungarian clinical laboratory system the word 'assistant' is equivalent to that of 'laboratory technician' in other countries.

Earlier there were two sorts of assistant training:

        a.   day-time course
        b.   evening course system

Candidates of this course have to be in possession of certificates of final examination in secondary school (GCE or Matura). The students are employed by laboratories; they work there as 'assistants without special schooling', they have to spend four hours twice a week in the teaching room for getting the necessary theoretical knowledge. After two years they sit for a state examination. Having passed the exam they are authorized to work in a clinical chemistry, microbiology, or hematology laboratory. After three years and with a history of excellent work they are eligible for a specialist course. They have an option to specialize in clinical chemistry, microbiology, or hematology. Getting through the examination successfully would authorize the assistant to enter a higher salary category.

Possible future developments
It is to be expected that Hungary, with its tradition of outstanding research and development in the medical field, will advance rapidly in the speciality of clinical chemistry. The efforts of the Hungarian Society for Clinical Pathology and the government to further the training of laboratory specialists and workers will alleviate a present shortage of staff. With definitive education and job programs, it is expected that the economic aspects of laboratory work will also improve to a level commensurate with the importance of the subject.

National Institute of Clinical Laboratory
Every branch of medicine in Hungary has a co-ordinating institute

(National Institute of Surgery, Radiology, etc.) so the Ministry of Health in 1975 established the National Institute of Clinical Laboratory. It is the main advisory body of the Ministry and it is also used to convey advice and instructions to medical laboratories throughout the country. Manpower problems co-ordination of purchase and supply of costly equipment, and quality control, all at the national level, belongs to the domain of the institute. It has a 16-member 'Collegium', i.e. a body composed of heads of various laboratories experts in clinical chemistry, microbiology and hematology. Their advice is sought in every important matter; they have a regular meeting in every second month.

National societies, associations and journals

National societies

Every professional man, physician, engineer, biologist, etc., working in this field can join the Hungarian Society for Clinical Laboratory Diagnostic.

The leader of the society is the president who directs the affairs of the society with the aid of the secretary and a committee of about 20 members. The leadership and the committee is re-elected in every third year. They have to give an account of their work yearly at a membership meeting. The number of the members on the 1st June, 1969 was 345; on 1st January, 1976 it was about 500.

The scope of the society's duties is of several kinds:

(1) It is the safeguard organization for the laboratory experts.

(2) It acted as an advisor for the Ministry of Health in laboratory questions until January, 1975.

(3) It promotes the scientific work of the Hungarian clinical laboratories by organizing meetings, congresses, and possibilities for making known and discussing the results of research work.

Scientific meetings can be divided into two groups. Six to eight times a year conferences are held devoted to methodology. New developments are discussed by invited speakers and partly by anyone who adds his remarks. If the new method seems to be of value it is introduced in the routine laboratory work.

General assemblies are held twice a year, one in Budapest, one in a large provincial town. On the first day of these meetings generally one topic is discussed. On the second day reports on research from any kind of clinical laboratory work are presented. The popularity of these congresses is sufficient that half of the membership participates.

(4) The mission of the society is to promote the progress of the Hungarian clinical laboratories. The hospitals in Hungary are partly under district or municipal council authority, partly under that of the Ministry of Health. To assist the authorities, the society conducted a survey of the condition of laboratories. The purpose was to provide for the following future possibilities:

(a)   To increase the capacity of the training centers for medicotechnical assistance.

(b)   To organize up-to-date equipment for central laboratories in larger districts where complicated examinations could be performed, which are impracticable in smaller hospitals.

(c)   To ameliorate the wage-system for laboratory personnel, because at present it does not attract enough physicians, or for other laboratory experts.

(d)   To introduce regular teaching of laboratory diagnostics at the medical faculties of the universities, as a special compulsory subject.

(e)   To publish a scientific review. Though in Hungary the number of the regularly published medical periodicals is over 50, it is difficult to publish laboratory themes.

(f)   To develop the importance of standardization of routine methods, so that examinations should be made at least in larger and well-equipped laboratories with the same methods. The first collection of 'Proposed Methods', was published about ten years ago. The second edition appeared in 1975.

Besides the society mentioned above, the Association of Hungarian Chemists has also a biochemical section dealing mainly with scientific questions of biochemical research. As this cannot be separated from clinical chemistry, the membership of these two societies is almost the same and members take part frequently in each other's meetings.

Meetings and journals

Annual meetings with international participation have been held on such subjects as endocrinology, liver function, lipid metabolism, laboratory in the postoperative care and the pediatric laboratory work. The society has had since 1974 a periodical of its own entitled *Laboratory Diagnosis* published four times a year. It contains new methods or modification of clinicopathological studies. The medical weekly *Orvosi Hemlap* and the bi-monthly paper *Kiserletes orvostudomany* also have special headings for laboratory publications. Regional meetings without pre-fixed themes were held several times in 1970. A congress of this society with international participation was held first in Budapest in 1971. The second at Balatonfüred, Lake Balaton, Hungary. The subject of the congress was clinical enzymology. The third congress in 1974 at Pécs was dedicated to the theme of the clinical laboratory in nephrology.

For about ten years the Hungarian Health Ministry has organized so-called 'Central Institutes' to promote the development of the respective medical branches. The Central Institute for Laboratory Work started to work at the beginning of 1975. The head of this department is the chairman of the above-mentioned Postgraduate High School. His task is to survey and to promote the laboratory work of the whole country with the aid of a 'collegium'. The members of this staff—about 15 laboratory experts—are designated by the Ministry of Health. They are heads of the greater laboratories and accomplish this special commission in addition to their routine work. This institute is now the advisor of the Ministry of Health in laboratory questions. So the society for Clinical Pathology is now rather a scientific association, the organization of survey controls having been also assumed by the institute.

INDIA

General introduction

Laboratory investigations as adjuncts to the health services of the country are built almost exclusively around hospitals which are mostly maintained by the state. Private hospitals, private clinics, and private laboratories catering for the requirements of private practitioners constitute only a very tiny fraction of the total facilities available.

There is no professional organization, central or regional, which lays down standards for training or services; these are being determined largely by the diploma and degree-giving agencies like universities, governments or colleges. The country has more than 100 medical colleges which prepare students for the first (bachelor's) degree in medicine and surgery and for the postgraduate degrees and diplomas in the different subjects, both clinical and non-clinical. Most colleges admit 100 to 150 students each year and thus there are 500 to 600 undergraduate students at a time in each of them. Six to ten students are admitted each year for postgraduate specialization extending over two to three years in each of the twenty or more specializations. Each college has an attached hospital with a bed-strength of 700 to 1,000. Large cities like Calcutta, Bombay, Delhi and Madras have three to five medical colleges and each college has its own hospital. The remaining colleges are spread more or less uniformly throughout the country, each with its own hospital. Hospitals somewhat smaller than the teaching hospitals and with a bed strength of 200 to 400 are located at the headquarter towns of all districts. These are called district hospitals and number about 200. There are several hundred still smaller hospitals scattered throughout the country with a bed strength of anything between 20 and 100. The clinical chemistry services in the country are organized more or less around the hospitals described above to cater to their needs.

Training

There is no uniformity between the different regions of the country in the matter of instruction in clinical chemistry. In certain well-endowed institutions like the All India Institute of Medical Sciences at New Delhi and the Postgraduate Medical Institute at Chandigarh in North India there is a three-year degree course leading to the degree of BSc or BSc (honours) in medical laboratory technology. Similar courses are also offered at some other centers. The course generally includes subjects like hematology, histopathology, microbiology and biochemistry, and instruction is given by the teaching staff of the respective departments.

Medical institutions located in several cities like Bombay, Calcutta, Delhi and Madras conduct a one-year certificate course in medical laboratory practice for matriculates in science.

In each medical college in the country all the medical students have to undergo a course in clinical biochemistry, which includes both theory and practice, as part of the undergraduate medical curriculum. The course covers the basic tests on urine, blood, faeces, cerebrospinal fluid and gastric juices. All candidates for the degree of MSc in medical biochemistry (mainly non-medical science graduates) and MD in biochemistry (all medical graduates) have to go through a compulsory course in advanced clinical biochemistry, which is largely modelled on the UK and USA pattern. This course is given by the senior teachers of the biochemistry teaching department. Candidates who aspire to the degree of PhD in biochemistry (both medical and pure science graduates can work for this degree) are also encouraged to go through this course, but it is not compulsory for them. Incidentally it may be mentioned that in about a third of the medical colleges in the country, the biochemistry departments are headed by non-medical biochemists. Though they usually get their doctorate degree by working on a pure research problem, they are required to become familiar with all the routine biochemical tests, both qualitative and quantitative, before they embark on their research problem, and they subsequently build up their experience and expertise in advanced clinical biochemistry to a high degree within a short period.

The Andhra University in South India has a two-year graduate course for chemistry graduates leading to the degree of MSc in the chemistry and microscopy of foods, drugs and water. Clinical biochemistry constitutes about one third of the whole course.

Services

Simple clinical investigations like urinary and blood sugar, blood urea and serum cholesterol are possible in all hospitals at the district level and above. The work is largely carried out by technicians who do not have a degree and is supervized by a biochemist, when one is available, or by the medical officer-in-charge. Specialized investigations in clinical chemistry are carried out in teaching hospitals in the biochemistry teaching department under the supervision of the professor of biochemistry, or in special multidiscipline laboratories under the technical supervision of the biochemistry staff. In metropolitan city hospitals clinical chemistry facilities are available round the clock.

Special investigations possible in major institutions include serum protein-bound iodine, serum enzymes like acid phosphatase, alkaline phosphatase, glutamic-oxaloacetic transaminase, creatine phospho-kinase, glucose-6-phosphate-dehydrogenase, urinary catecholamines, urinary 17-keto-steroids, 5-hydroxy-tryptamine in blood, corticosteroids in blood and urine, sodium and potassium in blood and urine, radio-active iodine uptake, etc. Still more specialized investigations are possible in a few advanced or upgraded departments of medicine, surgery, cardiology and the like, which have built up their own sophisticated biochemistry services as part of the speciality itself.

Instrumentation

Most laboratories are provided with microscopes, colorimeters, incubators and centrifuges. Several institutions have pH meters, thermostats and spectrophotometers, while a few have flame photometers, fluorescence spectrophotometers, instruments for measuring radioactivity and refrigerated centrifuges. Investigations like BMR and blood gas analysis are carried out generally in the physiology departments.

Automated analysis, which is now fairly common in advanced countries, has not come into vogue in India for several reasons. Equipment itself is very costly and involves considerable foreign exchange which is always in short supply. Even the maintenance of equipment poses serious problems; accessories, replacements and special chemicals need considerable foreign exchange. The country has not yet developed the required expertise in the electronics of sophisticated instruments, and when anything breaks down it may be months before an expert with the required knowledge arrives either within the country itself or from abroad. The country has a large number of men and women with moderate skills for whom employment has to be found. Though automated analysis has several advantages like small sample volume required for investigations, reliability of the results and quickness of the whole process, the cost factor militates very much against its widespread use in the country of low *per capita* income.

Conclusions

In India clinical chemistry as a reliable aid in the evaluation of the health status of individuals is yet to get the recognition that it deserves, in concrete terms. Facilities have to be created for more numerous and better laboratory investigations in order to enable correct diagnosis and regional treatment on a large scale. Much remains to be done to organize good laboratories, to provide them with equipment of the required quality and in the required numbers, and to staff them with well-trained personnel. There is a good case for doing away with the elementary certificate courses which do not attract students of a high calibre and which attempt to cover too wide an area rather thinly in too short a time. Professors of biochemistry who are at present the custodians of clinical chemistry in the country have been doing all they can to convince the administrators of the need for all the above-mentioned facilities, and to prod them to provide the necessary finances, but unfortunately all their valiant efforts flounder on the single factor of inadequacy of financial resources. There is certainly a great need for some form of standardization of the training of personnel and equipment for laboratories, but this may take a long time to be achieved in this large subcontinent with vast regional disparities in material and human resources.

IRELAND

Clinical chemistry as an independent discipline is of comparatively recent origin in Ireland. Indeed the first clinical biochemists were appointed to hospitals only in the 1950s. Prior to this the biochemical requirements of clinicians had been dealt with in university departments. Before dealing with the education and training of clinical biochemists it is useful to consider the development of biochemistry in Ireland, since most of the hospital biochemists take their initial degree in biochemistry.

Development of biochemistry in Ireland
The first department of biochemistry in Ireland was founded in Queens University, Belfast in 1924 under Professor Milroy. Later chairs were established in University College, Dublin, (Professor E. J. Conway, 1932), Trinity College, Dublin, (Professor W. R. Fearon, 1934), University College, Cork, (Professor R. J. Drumm, 1946) and University College, Galway, (Professor O. Ehocha, 1963). Most of these initial appointees were concerned with physiological and analytical chemistry yet could not have been considered clinical chemists as we know them today. They were mainly concerned with teaching medical and science students and research. However, much of the research has been on microanalytical methods and physiological chemistry and consequently was of benefit to those who later chose a career in clinical biochemistry.

Development of clinical biochemistry in Ireland
While this report is primarily concerned with clinical biochemistry in the Republic of Ireland, a very active and internationally recognized group of clinical biochemists has long been established in Northern Ireland. While their administration and organization is part of the UK system good liaison has always existed between clinical biochemists in the North and South of Ireland and is expected to continue to grow.

Early in the 1950s graduate biochemists were first appointed in charge of the biochemistry sections of pathology laboratories at the principal hospitals in Dublin, Cork and Limerick. These biochemists were science graduates without any formal training in clinical biochemistry.

The subsequent development of advanced biochemical techniques, expansion of the Health Service throughout the state and the increased demands made on the existing biochemical service has led to an increase in the number of laboratories, the extension of those already set up and the employment of many clinical biochemists.

Formation of the Association of Clinical Biochemists in Ireland

In the early 1960s it became clear that there was a need for an Association of Clinical Biochemists in Ireland to deal with scientific and professional affairs. In 1964 an association was formally established and a constitution drawn up. The structure and aims of the association were largely similar to those of the Association of Clinical Biochemists in Britain. Many of the clinical biochemists in Ireland at that time were members of the ACB in Britain and the close links established between the Irish clinical biochemists and their British colleagues in those days have been maintained ever since. Irish clinical biochemists regularly attend meetings and conferences of the ACB in Britain.

Structure of clinical biochemical services in Ireland

The clinical biochemistry service as it exists at the moment operates at several different levels:

(a) There is a small number of highly automated laboratories offering a centralized service; one of them also caters for seven other hospitals, and two others operate at a regional level and serve hospitals within a radius of 40 miles.

(b) Many (approx. 20) small laboratories carrying out a range of approximately 15 to 20 clinical biochemical tests.

ITALY

History

Up until 20 years ago clinical chemistry in Italy was regarded as an ancillary medical discipline and was considered an integral part of clinical laboratory technology, which included histology and pathological anatomy, microbiology, virology, hematology and clinical chemistry.

A clinical pathologist ('laboratory physician') with an almost exclusively medical background, was responsible for the laboratory. In fact, the only function of the clinical laboratory was for the assistance of physicians who requested analyses for their diagnosis. The greater part of the practical routine activity was delegated to the technical staff. At that time only technical aides were available, since the training of medical technicians had not been established.

The most important evolution for clinical chemistry in the laboratory started in the hospitals. A centralized laboratory service, at first only required in the larger hospitals, later became part of the common usage in nearly all.

Until the Hospital Law of 1968 the existence of laboratories dedicated to clinical chemistry alone was considered exceptional (in fact information regarding only three such laboratories has been found).

Initially the graduate staff were exclusively physicians, and no particular specialization was required. Certain laboratories did, in effect, use the service of a chemist who covered the need for skilled personnel in clinical chemistry, but without his ever being definitely and officially made part of the laboratory staff, or being able to participate in the prevailing career structures. This was because the norms governing hospital staff and organization had been laid down in the Hospital Law of 1938.

Only with the Hospital Law of 1968 was the official position of non-medical graduates in the hospital laboratory officially recognized and various career structures drawn up. The most important activity in the field of clinical chemistry was gradually given to graduates in chemistry and then to graduates in biological science.

The training of technical staff for laboratory work has developed slowly over the last 20 years. This has been due, above all, to the necessity for a laboratory routine extended to cover many different kinds of analyses and methods.

Having once realized the necessity for specific training, the major hospitals set up schools for technicians for their own medical laboratories. The character, duration and level of the courses varied a great deal from hospital to hospital, lasting from a minimum of six months to a maximum of three years. Admission was also variable. Some courses requested a diploma of a primary school level while others demanded a diploma from a pre-university level secondary school. The diploma given by the schools at the end of the training courses had no recognized legal value.

With the Hospital Law of 1968 the situation of these schools was brought clearly to light and the Health Ministry, which is responsible for their authorization started to lead hospitals towards a unification of courses, duration, admission requirements, etc. Given the multiple character of most of the hospital laboratories, however, it is obvious that the programs in the schools are also aimed at a multidiscipline education for the technicians. The most-appreciated technicians for clinical chemistry are those with a professional diploma in chemistry from a higher middle school and for whom attendance at a hospital school has represented only the completion of their analytical preparation concerned with the more medical aspects of laboratory work. A limited number of schools for biological or medical laboratory technology connected with the faculties of medicine have been in existence for many years. The highly qualified technologists graduating from these schools are very much in demand for research.

Since the laboratory has been considered one of the many spheres of professional activity for physicians, many medical doctors, general practitioners or hospital laboratory physicians have set up and invested in private laboratories. In recent years private laboratories conducted by non-medical graduates have also been opened.

114

The laboratory situation at present
In Italy the services for the carrying out of clinical chemical analysis are found as part of:
(a) hospital laboratories
(b) the laboratories of the university institutes within the faculties of medicine
(c) private laboratories
(d) other laboratories

In addition, some clinical laboratory activity, including clinical chemistry, is carried out in the laboratories belonging to the peripheral services of the Public Health Organization.

Hospital laboratories
The hospital laboratories serving the community amount to about 1,000, and of this number about 250 belong to regional or provincial hospitals, where specialized medical care is centered.

It may be calculated that the hospital laboratories which are adequately outfitted with the minimum amount of equipment needed to carry out routine analyses, form only two-thirds of the total number. In fact, measures are under way, wherever possible, to improve conditions and general equipment so that more laboratories will be in a better position to carry out the ever-increasing number of specialized diagnostic analyses and examinations.

At the time of writing this phase may be considered one of transformation, so that it is possible to find laboratories that are insufficient for the demands placed upon them next to laboratories that have already, for many years, been completely automated to cover a large workload and a wide range of different analyses.

The hospitals in Italy are independent of each other and so it is obvious that apart from a few exceptions, there is little collaboration for carrying out the more complex or rare analyses.

Laboratories in university clinical institutes
The situation in the university clinical institutes is completely different from that in the hospitals. Clinical chemistry or clinical biochemistry institutes still do not exist in the faculties of medicine, nor are there university central laboratories. The only possibility for a centralized laboratory service is for agreement between the university and a hospital administration. This means that the laboratory activity is often limited to small units in the various clinical institutes which are mostly interested in specialized analyses and sometimes in research. These units are under the direct responsibility of a clinical physician.

Well-defined structures do not exist and career possibilities for the staff (if they are not physicians) are extremely limited. In fact, biologists, chemists and pharmacists are employed as graduate technicians.

Private laboratories

The number of independent private laboratories in Italy cannot be given with precision. In 1965 the Health Ministry estimated that there were more than 6,000. The activity carried out by private laboratories is principally business-orientated and is closely tied to the various private health assistance societies.

Private laboratories with sufficiently complete and modern instrumentation and with well-trained staff do exist, but there are probably only a few hundred at this level.

On the whole they are small and their equipment is usually limited, so that in practice specialized analyses are not carried out directly in the laboratory itself but are passed on to other larger, better equipped laboratories.

Other laboratories

During recent years private outpatient medical centers have been set up and include clinical laboratory services. These laboratories are usually larger than the independent private laboratories mentioned above and automation and computers have been introduced in many already.

There are also many private hospitals in Italy but of these only the most important have their own laboratories; the greater number rely almost exclusively upon private laboratories.

The private health assistance societies also rely, for the most part, upon the private laboratories with which they may have an agreement. Some of them also have their own network of laboratories. The qualitative level of these laboratories on the whole is considered somewhat unsatisfactory.

Legal status

Italy is divided into 20 administrative regions and each region is then subdivided into provinces. Until recently the provincial medical authorities depended directly upon the Health Ministry in Rome which has now decentralized the organizations responsibility for hospitals and for health services to the regional health offices.

Hospital laboratories

The Italian Hospital Law number 132 (1968), substituting that established in 1938, has decreed that the laboratory service for clinical chemistry analysis and microbiology in the hospital is to be separate from that for histology and pathological anatomy. In the regional and provincial hospitals the clinical chemistry and microbiology service is divided into various sections (See Table 1). In hospitals with more than 900 beds two laboratories are set up; in hospitals with more than 1,800 beds, three laboratories are set up. In these cases the regulations provide for the distribution of the sections between the various laboratories.

116

It is for this reason that a great number of laboratories specializing in clinical chemistry have come into being in the bigger hospitals.

According to the law the graduate staff of these laboratories consist of medical doctors, biologists, chemists and other graduates with degrees in comparable subjects.

The organizational responsibility and career structure in the hospital laboratories are regulated by appropriate rules (See Table 2). In 1974 the new rules were still awaiting application in some hospitals.

Private laboratories

The authorization of the provincial medical officer, with the approval of the provincial health council, is necessary before permission to open a private clinical analysis laboratory may be granted. This rule was established by law in 1928.

Until a few years ago, the direction of private laboratories was considered the exclusive responsibility of a medical doctor; today as a consequence of the law of 1967 concerning 'the profession of biologist', the same possibility has been given to graduates who are inscribed into the professional role of biologists (apart from graduates with a degree in biological science, graduates in other disciplines fulfilling particular requirements can also be inscribed into this role). However, this argument has been under recent discussion and the new situation has still not been accepted completely.

Education and training for clinical chemistry

University courses

In 1967 the specific teaching of clinical chemistry was introduced for the first time as an optional course (entitled 'analytical clinical chemistry') for chemistry students at the institutes of analytical chemistry in the science faculties of the Universities of Rome and Naples.

Optional courses entitled 'biological analyses' or 'laboratory analyses' or 'applied biochemistry' and therefore, only in part dedicated to clinical chemistry are actually available for students of biology, pharmacy and medicine in various universities.

Postgraduate schools

For many years now the faculties of medicine have begun to set up schools of specialization for 'laboratory physicians'. In these schools clinical chemistry is taught but microbiology and hematology predominate. Specialized schools for laboratory physicians exist in the medical faculties of many universities—as for example at Bologna, Ferrara, Naples, Parma, Pavia, Roma and others. A degree in medicine is required for entrance into these schools.

Recently schools specializing in clinical chemistry available for non-medical graduates have been instituted. The non-medical graduates have in fact for many years been insisting upon the necessity of such courses in order to have the necessary qualifications for laboratory activity.

Workshops and seminars

Short courses on the theoretical and practical aspects of clinical chemistry or biochemistry for the training and updating of graduates already working in laboratories are held periodically in the larger hospitals and the university institutes. These courses are organized in the form of workshops or seminars and last from one to two weeks.

Training of laboratory technologists

Under prevailing Italian law, medical laboratory technicians are those who:

(a) have followed hospital courses of various lengths and with differing entry requirements (now only three-year courses for people with a diploma from the lower middle schools are authorized);

(b) have a higher middle school chemical technology diploma which has been followed by a practical course in a hospital;

(c) have a diploma from the special schools linked to the Universities of Milan and Padua. These courses last for two years and entry requirement is a diploma from a higher middle school.

The qualifications obtained correspond respectively to level C, B and A indicated by the World Health Organization.

Scientific societies

Società Italiana di Biochimica Clinica [SIBioC]

The SIBioC (Italian Society of Clinical Biochemistry) was organized in 1969 and is a member of the IFCC. Within this organization as an original nucleus, there is the Italian Association of Clinical Chemists, which was founded in 1968 and was based on professional interests. The SIBioC has an interdisciplinary basis and is made up of graduates in medicine, chemistry, biological sciences, pharmacy and other subjects dealing with clinical biochemistry (ordinary members), laboratory technologists and other culturally interested persons (i.e. students) who may also participate in the society's activities (associate members) and research institutes and societies (both private and state-owned) who are interested in the development of clinical biochemistry (collective members).

The statute of the society defines its activity as scientific, technical and cultural and promotes initiatives which attempt to improve the cultural preparation, the necessary professional updating and laboratory organization within the field of clinical biochemistry.

During the first five years of activity the society has organized three national congresses for clinical biochemistry and numerous other scientific meetings at national or regional level.

In 1974 there were about 500 ordinary members.

Società Italiana di Patologia Clinica [SoIPaC]
The SoIPaC (The Italian Society for Clinical Pathology) was founded soon after the SIBioC began its activity, and ordinary membership is granted only to graduates in medicine. This society stems directly from the Associazione Italiana Medici Analisti e Patologi (Italian Association for Medical Analysts and Pathologists) founded many years before and which even today bears traces of its original aims—the tutelage of the subject.

The society organizes an annual national congress to discuss scientific and technical subjects in clinical pathology and in addition, professional and union matters.

Comitato Italiano per las Standardizzazione dei Metodi Ematologici e di Laboratorio [CISMEL]
In addition to the above-mentioned societies there is also in Italy a committee for the Standardization of Methods in Hematology and in Laboratories (CISMEL) which, to date, has been predominantly interested in the standardization of hematology methods: according to its statute CISMEL is a closed shop.

Other societies
Numerous other societies and associations have an interest in clinical chemistry laboratory activities and among these there are:
  (a)   Società Italiana di Patologia
  (b)   Società Italiana di Ematologia
  (c)   Società Italiana di Endocrinologia
  (d)   Società Italiana di Medicina Nucleare
  (e)   Associazione Nazionale Italiana per l'Automazione

Possible future developments

Need of development
The need for a further and more complete development of clinical chemistry and its application has finally been acknowledged in Italy. Furthermore increasing importance is being given to the public health assistance services. Since 1975 changes have been going on in the entire field of health assistance administration and the institution of a National Health Service is now under study.

During the next five years the first concrete results in the following spheres should be seen.

The role of the university
General development will obviously be subject to a greater interest on the part of the university faculties in the creation of chairs for the specialized teaching of clinical chemistry and biochemistry, in university courses, and in the institution of new postgraduate schools for specialization.

119

The role of the health administration

Since at the moment of writing the change in the direction of health assistance from a central administration to a peripheral regional administration is still in process, it is obvious that some time will be needed to resolve all the present problems. However, the preparation of special regulations for clinical laboratories by some regional health administration is already under way.

The role of the Istituto Superiore di Sanità

The need for standardization in clinical chemistry is included among the new matters of interest for the Istituto Superiore di Sanità (the Italian National Health Institute).

In general this organization has the responsibility for carrying out technical co-ordination on a national scale. Moreover its activity is connected with European and international public bodies.

The role of the SIBioC

In future development the role of the Società Italiana di Biochimica Clinica will also be essential for the programming of future objectives.

Since 1969 the society has organized numerous interlaboratory quality control programs and has already begun a series of initiatives in favor of specialized teaching and or regulation in the field of diagnostic materials. At present the publication of an Italian journal especially dedicated to clinical chemistry and biochemistry is under study, and the setting up of work-groups is already at an advanced stage to encourage these activities and to help promote others.

TABLE 1.   THE LABORATORY SERVICES IN GENERAL HOSPITALS*

Laboratory service	Hospital category		
	Regional	Provincial	Area
Analysis, clinical	yes with sections	yes with sections	yes
Chemistry and microbiology			
Histology and pathology	yes with sections	yes with sections	optional eventually tied to preceding
Virology	yes†	yes†	no

* according to Law no. DPR 128 (27.3.69)
† if the hospital has an infectious disease department

120

TABLE 2   THE LEGAL STATUS OF EMPLOYEES IN THE HOSPITAL LABORATORY*

Medical graduates	Non-medical graduates (chemists, biologists and others similar)
Head Physician directs the laboratory service	**Director** directs a section
Assistant to Head Physician may direct a section if in possession of requisite qualifications	**Collaborator**
Assistant	**Assistant**

*According to Law No. DPR 128 (27.3.69)

JAPAN

For the first time in Japan a central clinical laboratory including a clinical chemical division was established in the Osaka University Hospital. Four years later legislation was enacted which established the requirements for the education and training of persons authorized to conduct clinical laboratory testing. Today every general hospital has a clinical laboratory service.

There are no fixed provisions for the qualification of director of university clinical laboratories or those in other medical schools. In general the post is filled by professors of medicine or biochemistry or by a selection of physicians specializing in medicine or clinical pathology. These latter individuals serve at the level of associate professor or lecturer.

121

Almost all hospitals maintain their own laboratories. In addition laboratory services are provided by commercial laboratories and those belonging to community medical centers. Most laboratories carry out clinical chemistry, hematological and microbiological analyses. Approximately 2,000 laboratories, mostly directed by medical doctors, exist in Japan.

Technicians who perform clinical chemistry, bacteriological, histopathological, parasitology, serological and hematological tests must meet the legal qualifications of the law established in 1958. Their training is provided in special schools, some of which are affiliated to university hospitals. Entrance to the program requires graduation from a senior high school. After three years of education and training the candidates are required to pass the national board examination to qualify for certification. The certificate carries official recognition. Medical doctors, pharmacists and veterinarians are qualified for certification. There are no special postgraduate courses. It is to be expected that in the future the education program will be extended to four years analogous to other specialized programs in universities.

The Japanese Society of Clinical Chemistry has been organized to promote scientific and professional activities in this field. It has published the Proceedings of the Symposium on Chemical Physiology and Pathology since 1961.

MEXICO

History
Hematology, serology, bacteriology and parasitology as well as clinical chemistry have been practised in Mexico for about the last 50 years. At present there are approximately 4,000 private laboratories where personnel carry out routine examinations in all these specialized fields. In public hospitals there is usually a division among these areas. Specialists in the subject direct the activities in each laboratory field. In large hospitals the director is generally a physician.

Legal status
By-laws of August 20th, 1941 of the Sanitary Code (Book V) provide regulations for the registration and requirements of the operation of private clinical analytical laboratories. Regulations of March 20th, 1964 established the requirements for clinical laboratory equipment, materials and personnel. The director may be a medical doctor, chemist or pharmacist, but he must have had specific experience in clinical laboratory practice.

Education and training for directors and senior staff of hospital laboratories
Formal programs for the training of directors and senior professional

staff are not yet organized. The qualification of such individuals stems from their experience and self-training in the clinical laboratories. An effort is now under way at the University of Mexico to establish specialized postdegree specialist training in each of the fields of clinical chemistry, hematology, bacteriology, immunology and parasitology. In 1970 a master's degree program in clinical chemistry was established in the faculty of chemistry. The program is designed to prepare senior staff for the direction of clinical chemistry laboratories. To increase more rapidly the numbers of such qualified persons a program was instituted in 1971 for the training of clinical chemists in leading laboratories abroad. Thirteen spent six months in clinical chemistry laboratories in the USA and Switzerland under this program.

Place of clinical chemistry in hospital services
It has been mentioned that in the larger hospitals the clinical laboratories are divided according to the laboratory specialities and are generally under the overall direction of a physician. In such a laboratory the clinical chemistry section is generally under the direction of a trained chemist. In the smaller hospitals the entire clinical laboratory function is generally unified and may be under the direction of a physician, chemist, pharmacist or microbiologist.

Place of clinical chemistry in non-hospital services
About 4,000 private clinical analytical laboratories presently function in Mexico. Most of these are in smaller cities and towns. They are generally under the direction of a physician, chemist, pharmacist or microbiologist. Invariably they are multivalent in the sense that all speciality procedures are carried out by the same staff in the same location.

Place of clinical chemistry in medical education
Medical students follow a course in general biochemistry, one of pathological biochemistry and short courses of microbiology, parasitology and immunology. Clinical chemistry is not taught as a single subject.

Place of clinical chemistry in non-medical education
The program of the master's degree in clinical chemistry in the chemistry faculty has been mentioned previously. Students in chemistry, biochemistry, and human biology follow a three-year university program which also includes courses in hematology, parasitology, immunology and bacteriology.

Technologists, technicians and other laboratory workers
Non-university technical schools in Mexico provide clinical laboratory analytical training for graduates with three years of secondary education. Upon completion of this program a *Laboratorista* certificate is issued.

Possible future developments

At present in Mexico, there are about 1,000 clinical chemists at various levels of education and training. Of these about 700 work in laboratories of public hospitals, mostly in the four larger cities of the country. This is a consequence of a general trend toward centralization, where the means and facilities exist for its achievement.

These numbers are insufficient to meet the present and future needs of the country, especially in rural areas. In addition, the growing complexity of clinical chemical laboratory activities demands a high degree of education and training to provide the quality of performance which our physicians and population have begun to recognize and demand in our present programs for the extension of health service delivery to the entire country. As a consequence of these developments, we expect in the future the rapid expansion of education programs and training facilities to provide qualified directors from the ranks of physicians, chemists, pharmacists and microbiologists.

THE NETHERLANDS

History

Clinical chemists started to function in hospital laboratories in 1945.

The Dutch Society for Clinical Chemistry was founded in 1947 (NVKC) and became a section of the Royal Dutch Chemical Society (KNCV). Membership of the NVKC consisted of those who had an academic education (PhD), and were interested in clinical chemistry. To promote the education of specialized clinical chemists and to support the interest of practising clinical chemists, within the NVKC an autonomous organization was founded, the Group of Certified Clinical Chemists. The admission to the Group was delegated to an independent Clinical Chemists Registration Committee which was organized on the analogy of the Medical Specialists registration in the Netherlands. Several other committees were instituted and international contacts were initiated. As one of the first results in 1954 the First International Congress on Clinical Chemistry was organized in Amsterdam. As a result of this first international contact the next year, 1955, an international journal for clinical chemistry, *Clinica Chimica Acta [CCA]* was established; the first volume was issued in January, 1956. Dutch initiative in the following years led to the organization of the West European Symposia, the first of which was held in Amsterdam in 1957, followed by national scientific meetings, two to three times a year.

Standardization, being recognized early as a very important subject, led to a close co-operation with the National Institute for Public Health (RIV). For example, in 1958 investigations leading to an internationally accepted, standardized Hb-method, had already commenced. Beginning with the institution of the NVKC (especially around 1960), several committees worked on national regulation, education, qualification and training of laboratory technicians in the Netherlands.

Education

Until 1966 the education of medical technicians was taken care of by private organizations, as the examination and the issue of the certificates were in the hands of the KNCV. Since 1966 the education and examination has been gradually taken over by official Analisten-scholen, subsidized by the government. Due to its great experience, the NVKC was able to advise the authorities in the composition of this new official branch of publicly maintained schools. Simultaneously clinical chemistry was recognized as a quickly developing, independent entity in university education, resulting in professional chairs.

After six years of elementary school and five to six years of high school the path to professional acceptance as a clinical chemist begins with about six years of university education which may be based in the fields of chemistry, biochemistry or pharmacy. The candidate may then continue to complete a PhD degree. After completing the academic study, which gives the title: *doctorandus* (doctoral examination for the doctor's degree) it is possible to specialize as a clinical chemist. This last phase must be completed in a qualified clinical chemistry hospital laboratory and the probation must be approved and is guided by the Registration Committee and lasts four years. After this period and after fulfilling all fixed requirements, the candidate is officially registered as a clinical chemist. To reach this goal several routes can be followed.

Doctor's degree in biochemistry

The probation time is three to four years in a qualified clinical chemistry hospital laboratory. This education comprises a program in anatomy, physiology, general pathology, hematology and microbiology, in addition to qualitative and quantitative clinical chemistry in all its aspects. Emphasis is placed on the clinical aspects of the profession. Clinical lectures and conferences must be followed, in order that the chemist can speak the same language as his medical colleague.

A registered clinical chemist in a qualified clinical chemistry hospital laboratory approved by the Registration Committee is paid for special aspects of the clinical chemistry laboratory such as management, automation, economy, etc.

Candidates from pharmacy and chemistry

These must fulfil the same requirements as mentioned above. In addition the probation time is four years and an examination in clinical chemistry, including physiology, physiological chemistry (i.e. the interpretation of the results of clinical chemistry investigation) must be passed.

Postdegree study
When a candidate has the doctor's degree in biochemistry or in a related field (at the discretion of the Registration Committee) the probation time may be shortened to 3—3½ years. Furthermore, the same requirements are valid as mentioned above.

Legal status

Negotiations during several years led in 1967 to recognition by the National Dutch Hospital Institute of the Registered Clinical Chemists. This was confirmed by the acceptance on both sides of a national basic hospital contract appointment. In consequence the NVKC (Dutch Society for Clinical Chemistry) was re-organized in the years 1969-1970, into a society with its own legal corporate capacity. It is still a section of the KNCV. (Royal Dutch Chemical Society). The Registration Committee, assisted by a visitation committee, functions as a completely independent entity with officially established qualifications, and rules, to qualify a 'certified clinical chemist'. At this moment the society has 450 members of whom 153 are also registered as qualified clinical chemists. These qualified clinical chemists serve 95% of the hospital beds in the Netherlands for clinical chemistry. A national foundation, 'Quality Control, The Netherlands', was established in 1973 in close collaboration with the NVKR.

NEW ZEALAND AND MALAYSIA

History

Clinical chemistry in New Zealand* was initiated at the University of Dunedin, Otago, under the leadership of Professors W. S. Roberts (1909-1916), S. T. Champtaloup and A. M. Drennan. An early application was made of the advances reported in the period 1912-1919 of the analysis of chemical constituents of blood. These were put into routine usage during the 1920s in the clinical laboratory. Additional new studies were carried out in iodine metabolism. At the same time the clinical chemistry laboratory at Christchurch Hospital became functional, followed shortly by related facilities at New Plymouth Hospital and the Wellington Hospital. From these beginnings the laboratories advanced in concord with the development of the field. Analogous activity in clinical chemistry was initiated in Auckland during the decade of the 1920s. By 1930 the staff had grown to seventeen in number, and all the routine determinations of the time were available for the physicians. From these beginnings the introduction of more sophisticated procedures and automation in the last decade have brought the laboratories to their present advanced status.

* Sims, F. H. (1969), 'Early hospital biochemistry in Australia', *Clin. Biochem.* 2, 405

In 1968 the clinical biochemists of New Zealand, approximately 70 in number, joined to form the New Zealand Association of Clinical Biochemists. This organization has now affiliated with the International Federation of Clinical Chemistry.

Legal status
While private individuals presumably could open laboratories in New Zealand, only those directed by a recognized pathologist and approved by the Health Department are allowed to claim any social security benefits. In New Zealand, under the National Health Act and the social security system a hospital biochemist is graded as hospital scientific officer and defined by government regulation as the holder of a degree in science from an approved university. The Health Act also recognized, however, medically-qualified practitioners as pathologists with the right to conduct their own laboratories and employ scientifically-trained staff.

There are no legal requirements for laboratory personnel in Malaysia, but it is doubtful whether persons without medical qualifications would be allowed to take blood from patients for laboratory tests.

Education and training for directors and senior staff of clinical chemistry laboratories
Medical graduates
All medical graduates received some training in clinical chemistry during their medical course. This is given however in most universities through the department of medicine or pathology. A professor of clinical biochemistry was appointed to the University of Otago, New Zealand. Other than the usual program in pathology there are no specific training schemes in New Zealand or Malaysia. There are no university courses in clinical biochemistry in New Zealand or Malaysia for science graduates. Those with a bachelor's degree in biochemistry or organic chemistry are appointed to positions in the large hospital laboratories where they receive their training from senior colleagues. Graduates in Singapore regularly travel to England for additional training.

Place of clinical chemistry in the hospital and non-hospital service
New Zealand
Medical services in New Zealand are controlled by the Minister of Health through the National Health Scheme which is similar to the British Health Service. They obtain their funds from compulsory contributions paid by the population with their income tax. All medical services in New Zealand are free, but the people are not debarred from engaging private doctors for fees.

The majority of the hospitals are large government hospitals, one of which is attached to the University of Dunedin. There are some small community or denominational hospitals. Large hospital laboratories are government institutions, completely controlled through the hospital

127

administration. They are directed by medically-qualified pathologists, but the biochemistry departments are under the control of science graduates with a large degree of independence. Some private pathologists have fairly large establishments employing scientifically-trained biochemists, a few private pathologists are also biochemists. Some private pathologists are at the same time in charge of a hospital laboratory. They obtain all their fees through the National Health Service.

Malaysia
While there is no national health scheme in Malaysia itself, Singapore has a comprehensive free medical health service similar to the British one for all those who are ill and need hospitalization. Treatment and tests are free in the Singapore hospitals.

There are two large laboratories in Malaysia, both under the control of their respective governments. One is the Institute of Medical Research in Kuala Lumpur which deals with all diagnostic laboratory tests for the area, though there may be very small laboratories in some hospitals. The biochemistry department of the Institute of Medical Research is under the direction of a PhD graduate (degree obtained abroad). He has some graduate staff with degrees in biochemistry from either the University of Singapore or the University of Adelaide, Australia. The technicians are trained in the laboratory. The laboratory of the Singapore General Hospital is under the administrative control of this hospital, which is the central government and teaching hospital. It is directed by a histologist with an MD degree and administratively under him is the director of biochemistry who has a PhD obtained abroad. He directs several graduates with biochemistry degrees from the University of Singapore who receive further training under his guidance, leading to higher degrees. The director has an honorary appointment at the University of Singapore and is also on the Scientific Advisory Council of the State. The technicians are trained in the laboratory. There are three small private clinical laboratories in Singapore doing only a very limited range of tests. One of these is directed by a bacteriologist, the second by a biochemist graduate from the Singapore University and the third by an ex-army officer.

Possible future developments
It is expected that New Zealand and perhaps also Malaysia may, because of a common background, follow the lead of Australia in regard to developments in clinical chemistry.

Societies, associations and journals
The New Zealand Association of Clinical Biochemists, formed in 1968 with approximately 70 members, is the only organization devoted entirely to the speciality.

NIGERIA

History
The term 'clinical chemistry' is relatively new in Nigeria. For about 40 years now, the name 'pathologist' has been attached to medical personnel who specialized in one or more of the following areas:

>Morbid anatomy
>Microbiology
>Hematology
>Serology
>Chemical pathology

As a result, clinical chemistry was not identified as a distinct discipline.

In recent years however, two groups of people, namely biochemists and physicians with a keen interest in biochemistry have worked in collaboration to stimulate renewed interest in the progress of medical science. It is now not uncommon practice for young doctors and biochemists with first degrees to strive to attain higher qualifications in clinical chemistry.

Legal status
The Federal Government of Nigeria promulgated a decree (no. 56) in 1968, establishing an Institute of Medical Laboratory Technology. Its primary duties are to determine the skills to be attained by those wishing to practise as medical technologists, laboratory technicians and laboratory assistants. The institute also maintains a register of all personnel, regulating their professional activities. The legislation does not define any criteria for the setting up of private medical laboratories except that the personnel employed to work in them must be registered with the institute.

In Nigeria, hospital services can be classified into three groups:

(1) University teaching hospitals
(2) State hospitals
(3) Private hospitals

There are six university teaching hospitals in Nigeria and each one is equipped with laboratories for bacteriology, clinical chemistry, hematology, parasitology and histology. The heads of these laboratories are usually medical doctors who have specialized in the appropriate area of study or occasionally non-medical doctors with a good training in laboratory medicine. Teaching hospital laboratories are bifunctional; they serve both as routine laboratories for the hospital and also as research centers. Each laboratory is staffed by a complement of medical doctors, biochemists, laboratory technologists, student laboratory technologists and laboratory assistants.

There are about 162 government-owned hospitals in Nigeria consisting of federal and state hospitals. (These do not include specialist hospitals like mental homes, leprosy care centers and infectious disease hospitals). Only about 50% of these hospitals have laboratory facilities and these laboratories are mainly suited for simple routine testing. The most common investigations carried out are microscopy, routine blood counts and chemistry and some blood grouping and cross-matching. Culture and serology are done in about 30% of these hospital laboratories. Each government-owned hospital laboratory is usually directed by a 'specialist pathologist' with a staff comprising technologists, assistant laboratory technologists, and laboratory assistants.

The private hospitals are very poorly equipped for laboratory analyses. Only about 14% of the 183 private hospitals in this country have laboratory facilities, and these are mostly missionary-owned hospitals. Activities in these laboratories, which are usually situated in one room, are confined only to the most routine procedures. Only six of these private hospital laboratories actually have full-time trained laboratory technologists. The others are manned by the medical doctors in charge of the hospitals aided by laboratory assistants.

Clinical chemistry in non-hospital services
In Nigeria there are only four private medical laboratories and they carry out investigations in hematology, chemistry, microbiology, parasitology and serology. These laboratories fill a minute part of the large vacuum created by the paucity of private hospital laboratories. There is a need for the setting-up of more private laboratories since the government-owned hospital laboratories cannot cope with the colossal amount of work they get. The directors of these laboratories are either medical doctors or biochemists who have held appointments with some of the university teaching hospitals.

Clinical chemistry in medical education
The teaching of clinical chemistry to medical students is incomplete. These students are taught biochemistry in their first two years in school and in the third year, they take a course in pathology. This includes clinical chemistry, hematology, microbiology and serology. They have very little laboratory experience besides routine urinalysis and blood counts.

Training of technologists and other laboratory workers
There are various categories of medical laboratory workers in Nigeria:

(1) Laboratory technologists
(2) Assistant laboratory technologists

(3) Student laboratory technologists
(4) Laboratory assistants
(5) Laboratory attendants

Most of the present day laboratory technologists in Nigeria, were trained in England where they obtained diplomas of the Institute of Medical Laboratory Technology. Their four-year training is initially broad-based to include hematology, clinical chemistry, bacteriology, parasitology, serology and in their last two years they specialize in one of these areas. In 1970, the Institute of Medical Laboratory Technology in Nigeria opened four schools for training technologists. These schools with one exception, are attached to the university teaching hospitals. Their training program is similar to that of the British Institute and the diplomas awarded are equivalent. The board of each school is made up of university professors and medical consultants. Successful secondary school graduates with good grades in mathematics, chemistry and biology may qualify to be student technologists. The teaching is well-organized and is modified to reflect peculiarities of our environmental conditions.

Clinical chemistry—present and future
The growing awareness of the importance of clinical chemistry in the understanding of clinical abnormalities has kindled a wave of interest in several young Nigerians. More and more doctors, biochemists and technologists are choosing to specialize in clinical chemistry. This enthusiasm in itself has created the problem of supply exceeding demand. There is a pressing need for the government to ensure that all government hospitals have their own laboratories and that these laboratories are staffed with qualified and competent personnel. Furthermore, the government should set up a body to advise all government hospital laboratories about advances in methodology, reagent sets, kits and instrumentation. This will not only reduce waste and save money, but it will also make for uniformity and comparability of results. Up to the present day, all hospital and non-hospital laboratories in Nigeria use manual methods only. This is necessitated by the relatively small volume of tests handled by each laboratory per day. A major setback for the laboratories is the difficulty in effecting repairs on damaged or malfunctioning equipment. All electronic equipment manufacturers are several thousand miles away from Nigeria and it usually takes about four months for spare parts to arrive there, and even longer when occasionally the whole equipment has to be sent back to the manufacturers for repairs. This problem is common to most parts of Africa. A solution would be to train more personnel in electronics, but it would be desirable also for the large manufacturers to maintain service stations on the African continent.

At the moment there are no standard laboratories or training schools primarily for clinical chemists in Nigeria.

Education
In 1975, the Nigerian Association of Clinical Chemistry was formally inaugurated. The principal aims of this association are:

(a) To promote education and research in clinical chemistry.

(b) To improve and maintain acceptable quality control in all clinical chemistry laboratories in Nigeria.

The association hopes, by its activities, to enhance the health care delivery system in Nigeria. This association is presently an affiliate of the International Federation of Clinical Chemistry.

NORWAY

History
The performance of laboratory services in clinical chemistry has been under the direction of fully qualified medical practitioners. In 1948 the Norwegian Medical Association accepted 'clinical chemistry and clinical physiology' as a separate speciality and formulated the requisite educational requirements. Starting in 1964 this was further divided to establish these previously joined specialities into individual entities. At the present time approximately thirty-five of the eighty somatic hospitals of Norway have created separate departments of clinical chemistry headed by a physician certified as a specialist in the field. The number of hospitals with separate departments of clinical chemistry is increasing. It is considered likely that about fifty of the larger Norwegian hospitals will have such departments within the next ten years. At the smaller hospitals, the clinical chemistry service is formally supervised by the head of the department of internal medicine which also legally is responsible for the work carried out.

A great need for clinical chemistry service arises from the work of the general practitioners. This need is for the most part covered by the outpatient service of the nearest hospital laboratory of clinical chemistry.

Societies
Several societies serve the interests of those specialists concerned with clinical chemistry. The Association of Clinical Chemistry and Clinical Physiology of the Norwegian Medical Association covers problems of professional and economic interests of its membership. The Society of Clinical Chemistry and Clinical Physiology is the scientific forum for clinical chemists of all types. The society is affiliated with similar societies in Denmark, Finland and Sweden to form The Scandinavian Society of Clinical Chemistry and Clinical Physiology. The regional journal is *The Scandinavian Journal of Clinical and Laboratory Investigation,* sponsored by The Scandinavian Society of Clinical Chemistry and Clinical Physiology.

Education—Clinical chemistry in the medical curriculum

At the four Norwegian medical schools (Tromso, Trondheim, Oslo and Bergen) there are at present five professorships and five positions as associate professors in clinical chemistry. In addition a few academic positions (professor or associate professor) have been founded in special branches of clinical chemistry (thrombotic research, surgical pathophysiology, neurochemistry and endocrinology).

Clinical chemistry and clinical physiology are compulsory topics in the curriculum of medical education. The medical students have to pass examinations in these topics, which carry the same weight as morbid anatomy. The teaching in clinical chemistry and clinical physiology is to a large extent carried out in connection with clinical pathological conferences, and in part the teaching is integrated with morbid anatomy.

Certified medical specialists in clinical chemistry

In addition to medical licensure the following is required for specialization in clinical chemistry.

Special training—Two years full service at an established department of clinical chemistry.

General training —

(1)   One year full service at a department of medicine or surgery or pediatrics or anesthetics. It is a requirement that an established department of clinical chemistry headed by a certified specialist in this field must be an integrated part of the hospital where this service is carried out.

(2)   Two years of work at a theoretical institution covering educational or research work in basic medicine or in chemistry. One year of this service may be replaced by service at a department of clinical chemistry.

(3)   At least six months of the complete education should be served at a department or laboratory actually engaged in diagnostic and/or therapeutic use of radioisotopes.

Non-certified physicians in clinical chemistry

There exist approximately 25 positions for MD trainees in clinical chemistry, attached to the 35 hospital departments of clinical chemistry in Norway. The greater number of these positions are occupied by MDs who intend to obtain certificates in clinical chemistry. Some positions are held by clinicians who include training in clinical chemistry in the education for other specialities, particularly in internal medicine, surgery and pediatrics.

133

Legal—delineation of 'clinical chemistry'

The rules and requirements for obtaining certification in the various medical specialities are worked out by the Norwegian Medical Association which also evaluates applicants and issues the certificate. In practice, the Norwegian Medical Association carries out the work by way of an affiliated 'specialist society'. The educational requirements for the speciality of 'clinical chemistry' define and limit the responsibility of the clinical chemists to the evaluation of the body status and organ functions based on the pathophysiological interpretation of chemical analysis. Thus, the speciality of 'clinical chemistry' is separate from specialities covering clinical microbiology, serology, immunohematology and morbid pathology. Nuclear medicine however, is usually attached to the clinical chemistry laboratory, although some functions may be attributed to the radiology department.

At the larger university hospitals the responsibilities of the clinical chemistry laboratory are practised largely in agreement with the said definition. However, at the majority of the smaller Norwegian hospitals the 'central laboratory' also takes on some work and responsibility in microbiology, immunohematology and clinical physiology.

POLAND

History

Laboratory diagnostics was officially organized as a separate speciality in 1953. At the same time the organization of large central laboratories started in contrast to previously existing small laboratories in each hospital ward. In 1953 the first state adviser for laboratory diagnostics was nominated by the Ministry of Health. In the following few years provincial advisers had also been nominated. Actually in each province there is an adviser in the field of laboratory diagnostics liaising with the local Health Service authorities. All provincial advisers, together with state advisers, form the Advisory Board representing laboratory diagnostics. The Advisory Board meets once a year to discuss the activity of each of the advisers in previous years and to decide what problems will be most important in the following year.

In 1962 the autonomous section of laboratory diagnostics in the Polish Medical Society was organized as the first representative body. After two years this section was transformed into the independent Polish Society of Laboratory Diagnostics (PTDL). In 1966 the first issue of the quarterly *Diagnostyka Laboratoryjna* (laboratory diagnostics), edited by the PTDL, appeared.

The main aims of PTDL activity is to contribute to the development of laboratory diagnostics in routine as well as in scientific aspects and to elevate the professional status of the laboratory workers.

PTDL has branches in all the provinces in Poland. There is also a possibility of organizing the section representing the subspecialities of laboratory diagnostics such as clinical chemistry, hematology, micro- biology, serology, parasitology, etc. To date one such section (namely clinical chemistry) has been organized. The full members of the Polish Society of Laboratory Diagnostics are laboratory workers graduated in science, pharmacy and medicine. Medical technicians can organize special, and to the same extent autonomous groups affiliated to the provincial branches of the society. Ten such technician groups are in existence.

Each of the provincial branches organizes monthly meetings which have a scientific and educational character.

Every three years, national congresses of the society are organized for the discussion of scientific problems and the election of the new authorities of the society.

The PTDL also publishes 'Information Bulletins' which are distributed to all members of the society and from time to time special monographs devoted to important problems, e.g. *Statistical Methods in Clinical Chemistry* by K. Grajnert.

Organization of laboratory diagnostics
In Poland in 1968, the Ministry of Health introduced general principles concerning the organization of laboratory diagnostics. On the basis of these principles the laboratories were divided into eight types of classes. This division depends on the location of the laboratory and on the range of its activity. There are the following classes:

(1) In outpatient health services there is a basic laboratory for about 20,000 inhabitants (for one part of a country district—in Poland the provinces are divided into districts); a basic laboratory in bigger towns for about 60,000 inhabitants; a district or quarter (in bigger towns) laboratory; and a provinical laboratory.

(2) In hospitals there is a laboratory in the smallest regional hospi- tals; a laboratory in district hospitals; a laboratory in town hospitals; and a laboratory in provincial hospitals.

Laboratories in specialized Health Service centers are classified separately. For each class of laboratories there are lists of the obligatory tests to be performed. The required professional qualifications for heads of the laboratories of different classes as well as for heads of each of the laboratory divisions (e.g. hematology, clinical chemistry and so on) are also stated.

The above-mentioned organization scheme is however, changing in conjunction with a general tendency toward integration of the whole Health Service. This is taking place by 'structural integration' of exis- ting regional laboratories into larger units required to perform hospital and outpatient laboratory services for an entire district or town quarter. Alternatively 'functional integration' is bringing about the centralized performance of some laboratory tests to avoid duplication of facilities.

This new organizational model is now being introduced all over Poland. When completed it will provide for the small district (peripheral) laboratories, the central district laboratories, and the great regional laboratories. An additional quite different form of laboratory diagnostics is performed in very small Health Service centers such as country health centers, ambulatory services in small factories, emergency stations etc. In 1965 the so-called, 'manual laboratory set' was introduced for routine use. This contains the necessary laboratory equipment and reagents in a removable box to enable the performance of some basic laboratory tests by non-trained persons (doctors, nurses, etc.) at any needed time. The list of these tests are as follows: blood-ESR, hematocrit, coagulation and bleeding time; urine—specific gravity, pH, protein, glucose, ketone bodies, blood, urobilinogen, bilirubin, feces—occult blood; gastric acidity. This manual laboratory set is now being modernized. The majority of methods are to be replaced by the strip tests and the range of methods will be broadened particularly with the possibility of the estimation of glucose and urea in blood, and the detection of significant bacteriuria.

Routine laboratories in Poland are obliged to perform tests using unified methods which are published in a special manual (first edition, 1958; second edition revised and changed in 1967). In 1967 the Scientific Advisory Board of the Ministry of Health appointed the Committee for Standardization of Laboratory Methods. The activity of this committee has the following aims: to choose the proper analytical methods; to check and standardize them, to elaborate a unified quality-control system to carry out the evaluation of laboratory equipment and diagnostic reagents produced in Poland or imported from abroad; to develop the system of statistics for laboratory diagnostics, and so on.

The Committee for Standardization publishes its recommendations one a year in a special issue of the journal *Diagnostyka Laboratoryjna,* which meanwhile has changed from quarterly into bimonthly.

In Poland the unified system of statistics in laboratory diagnostics has been in effect since 1968. The main principle of this system is the classification of all laboratory tests into five groups depending on the time required for their performance. Each test group possesses its own coefficient by which one must multiply the number of performed tests of this group. In this way the number of so-called laboratory units can be obtained which enables a more precise analysis of the workload and activity of each laboratory than is possible based upon the number of performed tests. The most helpful coefficients in the analysis of laboratory activity in this system are: (a) the workload coefficient that represents the number of laboratory units per one hour performed by a trained laboratory worker, (b) the structure coefficient which is the quotient of all performed units and all performed tests in the same period of time. This structure coefficient provides information about the proportion of difficult or more specialized tests performed by the laboratory, and (c) the last helpful coefficient is the so-called utilizing coefficient, which represents the number of performed tests per patient.

Education

Undergraduate education

During undergraduate education for medical students the program of laboratory diagnostics is rather elementary. There are only 45 hours of clinical chemistry in the third year of medical study. Almost all knowledge of laboratory diagnostics and especially the clinical aspects of these problems is supplied by clinicians. This unsatisfactory situation has a negative influence upon the development of specialization in clinical laboratory diagnosis. In 1970 special sections of clinical analysis were organized in pharmacological faculties of medical university schools. The educational program is provided during the last two years of study. This compensates for the lack of graduate staff in diagnostic laboratories and replaces the laboratory doctors in technical and methodological problems. It should be stressed, however, that the role of a laboratory doctor is as a trained consulting officer for the clinician rather than as a technologist performing laboratory tests himself.

Postgraduate education

Postgraduate education is carried out mainly in a specialized institution, the Medical Center for Postgraduate Education. In this center there is a department of laboratory diagnostics, the first scientific and educational center in the field of laboratory diagnostics in the country. This department organizes the courses for graduate laboratory workers and for candidates who intend to pass their examination for the II degree of specialization in the field. Apart from the department of laboratory diagnostics postgraduate education is also available in some centers all over the country. This peripheral activity is partially organized by the Medical Center for Postgraduate Education.

Three different kinds of courses are provided: (a) the basic course for doctors or other postgraduate persons beginning their work in a laboratory; (b) the advanced course for the above-mentioned candidates for II degree of specialization; and (c) the specialized courses (e.g. laboratory diagnosis of inborn errors of metabolism, laboratory diagnosis of protein metabolic disturbances, etc.). In Poland there are five centers in which the basic courses are given; two centers for advanced courses and eight centers for specialized courses. The number of these centers changes each year and depends on the actual needs. There is the special Programming Committee appointed each year by the Medical Center for Postgraduate Education which decides this matter.

The specialization in laboratory diagnosis is a function of the administrating authorities rather than the Polish Society of Laboratory Diagnostics. Specialization is available only for doctors and pharmacists, each having different names of the specialities and different programs.

Doctors are the specialists in 'laboratory diagnostics', a field in which the diagnostic aspect of training is stressed, while pharmacists are 'clinical analysts'. In both variants the time of training is the same: three years for the I degree and additionally two years for the II degree. During this time the candidate must be acquainted with all practical and theoretical problems included in specialization programs. Each candidate has his own tutor who must be a specialist with a II degree and who serves as the mentor during all the specialization schooling. After this time the candidate asks his provincial adviser for permission to pass the examination, and the provincial Health Service office to place him on the list of persons passing the examination. The examinations for the I degree are organized by the provincial Health Service offices, while for the II degree the examinations are organized centrally only in the department of laboratory diagnostics in the Medical Center for Postgraduate Education.

Research

Not long ago there was only one scientific center in the field of laboratory diagnostics, namely the above-mentioned department of laboratory diagnostics. Recently as a result of organization changes in the structure of medical university schools some departments of laboratory diagnostics with scientific, educational and routine programs have been added. This should stimulate the number as well as the quality of published scientific papers and also should encourage young doctors to choose laboratory diagnostics as a subject of their scientific and professional interest.

PORTUGAL

The organization of clinical chemistry or clinical analysis in Portugal depends upon the population, geographical location and the availability of university hospitals and faculties. In practically every small city and village of any significance there are private laboratories which belong to specialized medical doctors or sometimes to pharmacists. Small hospitals may have a 'private' contract with a specialized medical doctor or even with a pharmacist who does the analysis for the hospital patients. In cities where there are no university hospital laboratories, the laboratories are usually directed by a specialized medical doctor. Larger non-university hospitals should theoretically be able to accomplish all laboratory services required for their patients. However, advanced analytical services are not available. The direction of these laboratories is reserved to medical doctors; in addition, physicians serve as assistant directors and interns.

In university hospitals each service or institute of the medical school has its own private laboratory where routine and more common clinical analyses are carried on as well as specialized analysis which is of interest to the particular university center where it takes place.

Besides these 'small' laboratories there are the so-called central laboratories where the greater part of analysis demanded by clinics working at the hospital is concentrated.

Organization of laboratories

There are three sections or three central laboratories, their names indicating the orientation: Central Laboratory of Bacteriology, Central Laboratory of Hematology and Central Laboratory of Biochemistry.

Each of these sections is directed by a specialized medical doctor. There is also a director of laboratories, a medical doctor, who supervizes central laboratories in the same way as the 'advanced' ones of the various services. Only specialized medical doctors are allowed to do analysis. This title is given by the medical doctor's syndicate, after examination by a jury of specialists designated by the syndicate or after public examinations to fulfil official vacancies in state hospitals. The approval of these examinations is equivalent to those held before the jury of the syndicate.

In all hospitals and in some private laboratories there is auxiliary personnel for clinical analysis, mostly females, who have the title of 'preparers'. This is granted after a special course taken in official laboratories and a final examination.

Caixa de Previdênica [Social Fund Institution]

This is a state organization with special characteristics. The so-called 'Federation of Social Fund Institutions' has its own analysts who are paid for each analysis and who have a fixed salary. This Federation due to its large size, includes a great percentage of the Portuguese population.

The Federation of Social Fund Institutions is part of the Ministry of Corporations and Social Funds; the remaining sanitary organization in the country is part of the Ministry of Health and Assistance.

RUMANIA

Education and training for directors and senior staff of clinical chemistry laboratories

In Rumania both medical and non-medical graduates function in clinical chemistry laboratories.

Medical graduates

Clinical laboratory specialization is achieved in three stages:

(a) *First grade*—Periodic examinations lead to admission at this level of postgraduate qualification. The first-grade period extends over three years and consists of the following types of training:

(i)   Daily work in a clinical laboratory under appropriate supervision.

(ii)   Attendance at a special course of nine months at the postgraduate medical faculty. In the program of this course, basic practical and theoretical aspects of clinical chemistry are considered. Problems of theoretical and applied chemistry, not usually stressed in medical university training, are given special attention.

The curriculum for first-grade medical graduates consists of lectures and practical work extending over a three-month period.

The lecture program (six hours weekly) includes:

(1)   The basic concepts of analytical chemistry.

(2)   The chemical structure of living matter.

(3)   The theory of some physico-chemical methods of analysis; colorimetry, spectrophotometry, chromatography, electrophoresis.

(4)   Metabolic pathways related to human disorders.

(5)   The body electrolytes and their disturbances.

(6)   Enzymes and their properties.

The practical work (24 hours weekly) consists of:

(1)   The electrophoresis of serum proteins and lipoproteins.

(2)   The estimation of normal and pathological constituents of urine; proteins, glucose, chloride, phosphorus calcium, urea, uric acid, creatinine.

(3)   The estimation of blood glucose, blood urea, ammonia in blood, serum proteins, iron, phosphorus and calcium in serum, bilirubin, lipids, cholesterol in serum.

(4)   The estimation of serum enzymes; acid and alkaline phosphatases, aldolase, aminotransferases, ornithine carbamyltransferase, dehydrogenases.

(b)   *Second grade*—At the end of the three years of the first grade and on successful completion of an examination to become a 'specialist' the candidate may be employed in every clinical laboratory in a post with full responsibility. At maximum five-year intervals, the specialist is expected to attend a postgraduate course of three months, at the postgraduate department of biochemistry. During this course, the specialist reviews the theory and practice of the most recent analytical techniques, and discusses the actual problems of theoretical biochemistry.

(c)   *Primary grade*—After at least ten years spent in the second grade, specialists must pass an examination to obtain the primary grade. At this professional level he may be in charge of a clinical laboratory of a large sanitary unit. Medical graduates must attend postgraduate courses at maximal intervals of three to five years.

Non-medical graduates
In clinical chemistry laboratories there are also graduates of the faculties of chemistry and biology.

The various groups are as follows:

(a) *Basic grade*—All new graduates must spend at least one year in the first grade training in a laboratory of clinical chemistry under the supervision of a biochemist or a specialist.

(b) *Chemist of biologist grade*—(hospital biochemist) is attained at the end of the first grade year upon passing an examination to become a hospital biochemist. At two to four year intervals the biochemist must attend a two-month postgraduate course on 'new techniques' at the department of biochemistry of the postgraduate medical faculty.

The curriculum for the hospital biochemist consists of a two-month program of lectures and practical work.

Lectures (four hours weekly) encompass:—

(1) The theory of some modern physico-chemical methods of analysis: spectrophotometry, flame photometry, chromatography, gel filtration, immunoelectrophoresis, fluorimetry, high-voltage electrophoresis, ultracentrifugation.

(2) Metabolic pathways related to human disorders.

(3) Protein biosynthesis.

(4) The plasma proteins.

(5) Mechanism of enzyme action. Regulation of enzymatic pathways.

(6) Indications and clinical value of the enzymological investigations.

(7) Perspective and limitations of clinical enzymology.

(8) Liver functions; its biochemical disturbances and exploration.

(9) Renal function.

(10) Statistics and quality-control in clinical chemistry.

(11) Automation in the laboratory.

(12) Theoretical problems in clinical chemistry.

Practical classes

(1) Column chromatography and ion exchange chromatography.

(2) Thin layer chromatography.

(3) Gel filtration, immunoelectrophoresis.

(4) Ultramicro analysis.

(5) Various types of in vitro tissue preparations.

(6) The Warburg apparatus.

(7) Cell fractionation: preparation of mitochondria, lysosomes, microsomes; enzymes in subcellular fractions; determination of:-mitochondrial succinic dehydrogenase, glutamic dehydrogenase, micro-

somal glucose-6-phosphatase; enzymes of the soluble fraction; glucose-6-phosphate dehydrogenase, isocitric dehydrogenase, xanthine oxidase.

    (8)   The ultracentrifuge; the tissue enzymes.

    (9)   Estimation of isoenzymes.

  (10)   Determination of Michaelis constants.

  (11)   Simplified tests.

  (12)   Flame photometry of electrolytes.

(c) *Principal grade*—This status is reached after at least ten years' experience as a hospital biochemist and upon passing an examination to become principal biochemist in charge of the biochemical department of a large laboratory.

SOUTH AFRICA

History

Clinical chemistry in the Republic of South Africa is generally known as chemical pathology reflecting the British influence on the early development of medical practice in this country. The chemical pathology laboratories originally existed as branches within clinical pathology departments but the rapid development of clinical chemistry over the last two decades led to the establishment of separate departments. All six medical schools now have separate university departments of chemical pathology and the South African Medical and Dental Council recognizes chemical pathology as a registrable speciality for medical practitioners and medical technologists.

The profession is dominated by medically-trained chemical pathologists and it is only in recent years that increasing numbers of biochemists are being employed in university departments, and their appointments are usually to research posts rather than to clinical service laboratories. Such biochemists do not have a recognized title and are generally known as medical biochemists, clinical biochemists, research assistants or professional officers.

Legal status

In the Republic of South Africa the medical and allied professions are controlled by the Medical and Dental Council. All registered medical practitioners may provide or supervise clinical laboratory services but in addition medical practitioners whose postgraduate training and experience meet the requirements of the council may be registered as specialist chemical pathologists.

Similarly suitably qualified medical technologists are registered by the council through the Professional Medical Technology Board. In contrast science graduates engaged in clinical biochemistry have no legal status at all and many in fact seek registration as medical technologists.

Societies and associations
The following professional societies and associations cater for the profession of chemical pathology in South Africa.

The Medical Association of South Africa
The Medical Association is open to all registered medical practitioners including specialist pathologists and chemical pathologists. The pathology subgroup of this association is the only organization legally recognized by the state as representing pathologists in the country. Nonmedical biochemists are not eligible to join the Medical Association.

South African Society of Pathologists
The members of this society include microbiologists, anatomical pathologists, virologists, hematologists as well as chemical pathologists. Biologists, biophysicists and biochemists associated with departments of pathology are also eligible for membership. The total membership is in the region of 300 and the prime function of the society is to organize an annual scientific congress which it has been doing for the past ten years.

Society of Medical Laboratory Technologists of South Africa
This society is open to all medical laboratory technologists and has a membership of approximately 600. It is affiliated to the International Society of Medical Laboratory Technologists and holds regular regional and national meetings, and publishes a monthly journal.

South African Association of Clinical Biochemists
This organization came into being in February 1975 to represent the interests of both chemical pathologists and clinical biochemists and application has already been made for affiliation to the International Federation of Clinical Chemistry. The initial membership will be about 50 of which more than threequarters will be medical specialists.
The association will cater for the specific needs of chemical pathology (clinical chemistry) in the country and will hold scientific meetings in conjunction with the Society of Pathologists.

Chemical pathology services in South Africa
Pathology services, including chemical pathology are provided by five independent groups:

University medical schools
Departments of chemical pathology in the six medical schools at the universities of the Witwatersrand, Pretoria, Cape Town, Stellenbosch, the Orange Free State and Natal. These university departments are all attached to large provincial teaching hospitals for which they provide diagnostic service facilities in addition to their teaching commitments.

These laboratories are financed by the provincial administration concerned.

Laboratories of the South African Institute for Medical Research
The South African Institute for Medical Research, SAIMR, has a large laboratory complex in Johannesburg—which has now combined with the medical school of the University of the Witwatersrand—and numerous smaller laboratories throughout the country except in the province of Natal.

State Health Laboratory Service laboratories
The State Health Laboratory Service laboratories are largely responsible for public health services, toxicology and forensic pathology, but are taking over responsibility for routine clinical pathology services in a number of new state hospitals. There is no separate chemical pathology section within the service but laboratory planning includes separate specialized biochemical laboratories in the larger hospitals.

Provincial Laboratory Services
Only the provincial administration of Natal operates a separate laboratory service within the hospitals department. There is no separate chemical pathology service but separate facilities exist in the large laboratories.

Laboratory services for private patients
Pathologists in private practice work entirely in the larger urban complexes and about 20 laboratories provide clinical laboratory service for private patients.

The teaching of clinical biochemistry
All six medical schools and the SAIMR have separate independent departments of chemical pathology which are responsible for undergraduate and postgraduate medical education as well as the training of clinical biochemists and medical laboratory technologists.

Undergraduate medical training
There is no uniformity between the various universities but all offer a full undergraduate course in chemical pathology to medical students in the third, fourth or fifth year of study. Basic chemistry and biochemistry are provided in the preclinical years through the department of chemistry and physiology. Correlation between the basic preclinical courses and the course in chemical pathology is largely lacking but attempts are being made to improve the integration of the teaching of chemical pathology with the other clinical disciplines. The emphasis in tuition is on formal lectures with varying but lesser amounts of practical laboratory work, tutorials and clinicopathological case presentations.

144

Qualifying written, oral and practical examinations are held at the end of the course. Although there is no standard curriculum or examination the interchange of external examiners between universities contributes something towards maintaining a uniform standard of training throughout the country.

Postgraduate medical training
Postgraduate medical training in chemical pathology is offered at all six medical schools and the SAIMR. The minimum period of training is three years and candidates can then acquire a higher qualification recognized by the Medical and Dental Council in one of four ways:

(1) All the universities confer the degree of Master of Medicine in Pathology (MMed(Path)) on successful completion of written, practical and oral examinations; some also require a thesis in addition.

(2) The faculty of pathology of the South African College of Medicine offers a two-part examination and awards a Fellowship (FFPath) which is recognized by the Medical Council.

(3) It is also possible for candidates to sit overseas examinations which are recognized by the Medical Council, such as those offered by the Royal College of Pathologists in Britain.

(4) The Medical Council may also accept as a qualifying degree a doctorate (MD) obtained by thesis at a recognized university.

The postgraduate training offered differs in detail from one center to another but consists largely of in-service training in which pathology registrars serve in hospital laboratories. The amount of formal tuition offered is generally minimal.

At present most registrars in pathology seem to prefer to qualify and register as specialist clinical pathologists rather than as chemical pathologists since under the regulations of the Medical Council a clinical pathologist may practice in all branches of pathology whereas a chemical pathologist is barred from practising anything but chemical pathology.

Training of clinical biochemists
Undergraduate training takes place in normal university departments of chemistry or biochemistry and leads to the degree of BSc. These courses do not in any way prepare the biochemist for clinical work.

Several medical schools are now able to award the degrees of MSc or PhD to biochemists who do further postgraduate research towards a thesis in departments of chemical pathology. There is, however, no formal training towards clinical biochemistry and no clinical degree (such as the British master's degree in clinical biochemistry) is available. For the most part biochemists in the medical schools are predominantly involved in research although the shortage of suitably qualified medical graduates has led to more biochemists taking on supervisory duties in service laboratories.

145

Training of medical laboratory technologists

In contrast to clinical biochemists the training of medical laboratory technologists is organized on a national basis. The course covers four years. During the first two years students rotate through all pathology departments and are required to pass an intermediate examination in physics, chemistry, anatomy and physiology, and laboratory technique. The latter two years may be spent in various sections—in which case the technologist can qualify in clinical pathology—or the candidate may concentrate on chemical pathology and receive a diploma in this subject alone.

The training and examination of medical laboratory technologists is supervised by the Department of Education, Arts and Science in consultation with the Professional Medical Technology Board. In-service training is offered at all the medical schools as well as other approved state, provincial and private laboratories and six to twelve months is spent at a technical college. The department of health is not directly concerned with the training of pathologists, biochemists or technologists other than those training in state health laboratories.

Conclusion

A serious deficiency exists in the clinical biochemistry training in South Africa in that there is neither formal training nor accepted status for the clinical biochemist. Against this must be weighed the dearth of medically-trained applicants and the numerous vacancies which exist at all levels for professional staff in chemical pathology laboratories throughout the country.

The revision of training schedules with a view to rationalization and consolidation of courses at all centers together with concerted efforts to obtain recognition for the clinical biochemist are problems which should be an immediate concern of the Association of Clinical Biochemists.

SWEDEN

History

Until 1938 chemical investigations on specimens from patients were performed, with few exceptions, in hospital laboratories. The heads of the various hospital departments were responsible for small laboratory units of their own. As methods became more complicated and workloads increased the benefits of a central laboratory organization was realized. The first central laboratory with an independent head physician for clinical chemistry was built in 1938. The first professors for clinical chemistry were elected in 1956. By 1969 there were 43 central laboratories and about 100 medically-trained clinical chemists, 75 engineers, 200 laboratory nurses, 550 technicians, 400 laboratory aids, 100 clerks and 100 unqualified individuals. The growth rate has continued since then. The Swedish Society for Clinical Chemistry was founded in 1954.

The organization of laboratories

Sweden is divided into seven hospital regions. The 'regional hospital', as a rule, a university hospital of 1,000 to 1,500 beds, may also serve as a 'central hospital' for its county. Such a center may have 50 to 80 persons functioning in clinical chemistry laboratories. In addition other 'central hospitals' in a county of 400 to 700 beds provide services in defined geographic areas. The central hospitals have all the specialities except the more exclusive ones, e.g. neurosurgery and thoracic surgery. The smallest units are organized with departments of internal medicine, surgery and limited facilities for physiological and x-ray investigations and clinical chemistry. In both the regional hospitals and the central hospitals one organization including a central laboratory and often several clinical laboratories in the different departments is responsible for the chemical work for both inpatients and outpatients. This means that the clinical chemist is in charge of space, personnel and methods in the central laboratory and the clinical laboratories. There has been a trend during the last five years to use the peripheral clinical laboratories as blood sampling stations, for simple chemical tests and to centralize as much as possible to the central laboratory. The laboratory personnel is often used to take both capillary and venous blood samples on both inpatients and outpatients.

Small laboratory units in county hospitals of 50 to 200 beds often co-operate with the central laboratories. As clinical chemists are often directly responsible as at least consultants to these laboratories there has been a trend also to send the more complicated analyses to the central laboratory and keep those methods which are necessary for the immediate handling of outpatients and the emergency service. Clinical chemists are responsible for the chemical work including hematology in all the different types of hospitals except in the smallest units, where one of the clinical physicians is often in charge of the laboratory. Laboratory physicians including clinical chemists are as a rule, however, working as consultants in these hospitals.

Education

All medical students receive training in clinical chemistry during their preclinical period (i.e. during the first part of their third year). The course is given by the professor of clinical chemistry and lecturers in clinical chemistry and includes both lectures and practical training during 30 hours for each student. The departments of clinical chemistry at the university hospitals all have their own laboratory facilities for education.

The students also spend about two weeks in the laboratory during their course in internal medicine. Lectures are given on selected topics during the courses in internal medicine and surgery, often in collaboration with the respective departments.

Medical graduates who decide to become clinical chemists enter a university or central hospital laboratory.

The specialist in clinical chemistry has to follow the course of study prescribed by the national health authority. After obtaining a licentiate in medicine the postgraduate student spends at least one year in the medical clinics followed by a minimum of two years in theoretical medical biochemistry. While engaged in research in such a department he is also employed as a demonstrator teaching medical students. After his years in the theoretical departments, the candidate must spend a minimum of two years in a department of clinical chemistry. At the end of five years' training he is recognized by the government as professionally competent in clinical chemistry and is graded as a specialist. Usually it takes a minimum of eight years to become head of a department of clinical chemistry, and it is unusual to be appointed head of a department without an MD, DSc degree.

Legal

The pathway leading to specialization in clinical chemistry proceeds by way of a medical education as has been described above. Consequently almost without exception the head of the department will be a physician. Although no examination is given at the end of the prescribed training and education, only such qualified individuals are appointed to the posts.

SWITZERLAND

The Swiss Society of Clinical Chemistry is concerned with the education and training of clinical chemists in Switzerland. Candidates educated according to its guidelines and who have been awarded a certificate of special training in clinical chemistry are recommended to universities for appointment to posts in clinical chemistry.

The clinical chemist must be familiar with the usual methods of the clinical chemistry laboratory, be well acquainted with the general chemical and biochemical working procedures and be capable of handling problems of pathological biochemistry. He must have some basic knowledge of clinical medicine.

The career of a clinical chemist is open to medical graduates, chemists, biochemists and pharmacists after the completion of their studies (state examination or diploma). The special training has a duration of two and a half years (for biochemists and pharmacists: two years). It complements medical studies on the biochemical side, and chemical studies on the medical side. The progress of training is substantiated by certificates from the heads of qualified laboratories in which it occurs. The first part (approximately one year) consists of lectures and practicals. Its program is as follows:

1. Program for medical graduates with a state diploma

[a] *Lecture courses*		*Hours*
(1)	Analytical chemistry (including knowledge of instruments)	50
(2)	Selected chapters of physical chemistry	50
(3)	Complementary teaching in mathematics, statistics and quality control	40
(4)	Special organic chemistry and biochemistry	80
(5)	Electronic data processing	70
(6)	Clinical chemistry (if this course exists, it can replace part of 1, 2 and 4.	

[b] *Practicals [760 hours]*		
(7)	Organic chemistry	1 semester
(8)	Analytical chemistry	½ semester
(9)	Physical chemistry	½ semester
(10)	Exercises in mathematics	40

2. For chemists with a diploma

[a] *Lecture courses*		*Hours*
(1)	Appropriate parts of general physiology and physiopathology	140
(2)	Clinical chemistry, general and clinical biochemistry	140
(3)	Appropriate chapters of general pathology	105
(4)	Optional among: immunology, bacteriology, hematology, serology	35
(5)	Statistics and quality control	35
(6)	Electronic data processing	70
(7)	Human genetics	20

[b] *Practical and visits*
These consist of 500 hours to be distributed optionally among the following branches: urine analysis, hematology, serology, bacteriology, histology, cytology, endocrinology.

3. For graduate biochemists

[a] *Lecture courses*		*Hours*
(1)	Appropriate chapters of general physiology and pathology	140

149

(2)	Clinical chemistry and clinical biochemistry (minimum)	70
(3)	Appropriate chapters of general pathology	105
(4)	Optional among: immunology, medical bacteriology, hematology, serology	35
(5)	Quality control	15
(6)	Electronic data processing	70

[b] Practical and visits
They consist of 350 hours to be distributed (optionally) among the following branches: urine analysis, hematology, serology, bacteriology, histology, cytology, endocrinology.

4. For pharmacists

[a] Lecture courses		*Hours*
(1)	Appropriate parts of general physiology and physiopathology	140
(2)	Clinical chemistry, general and clinical biochemistry	110
(3)	Appropriate chapters of general pathology	105
(4)	Optional among: immunology, bacteriology, hematology, serology	35
(5)	Statistics and quality control	35
(6)	Electronic data processing	70
(7)	Human genetics	20
(8)	Selected chapters of physical chemistry	50

For (1), (3), (4), (6) and (7), courses followed during the studies can be included.

Practicals
They consist of 500 hours to be chosen among the following branches: urine analysis, hematology, serology, bacteriology, histology, cytology, endocrinology. Practicals during other studies can be incorporated.

The remainder of the training time consists in practical activity in one or several qualified institutes or laboratories. At least two-thirds of this activity should be devoted to a chemical branch. The rest is preferably accomplished in a laboratory dealing with morphological analyses.

It is recommended that candidates submit a scheme of planned studies to the Swiss Society of Clinical Chemistry before beginning their training.

Those who successfully complete their specialized training will receive a certificate from the Swiss Society of Clinical Chemistry testifying that they are qualified for directing a laboratory of clinical chemistry. This decision is taken by the Committee of the Swiss Society of Clinical Chemistry.

SYRIA

History
A few hospital laboratories covering microbiology, chemistry and hematology were established as separate units headed by MDs or pharmacists in 1919. (The Damascus University, then Syrian University, the Ministry of Health and the Army). After a year a unit for performing toxicological tests was created in the Ministry of Health. These laboratories developed slowly until 1953 when a section of clinical pathology was created in the university hospital, directed by a medical doctor, specialist in laboratory examinations. This section became part of the department of pathology in 1958. In 1971 the independent department of laboratory medicine was created, and developed rapidly, including for the time being an important section of clinical chemistry, as well as sections of hematology, blood banking, immunology and microbiology.

Legal status
Medical doctors or pharmacists with a specified postgraduate training acquire the status of 'specialist' in one of laboratory medicine branches, including chemistry, by application to the Ministry of Health and upon presentation of their training diploma.

Education and training for directors and senior staff of clinical chemistry laboratories
There are two ways to obtain a certificate accepted by the Syrian Ministry of Health in this speciality:

1. The diploma of higher studies in laboratory medicine, delivered by the faculty of medicine after a three-year course given to medical graduates. The program includes theoretical as well as practical courses in clinical chemistry, analytical and physical chemistry, statistics, microbiology, hematology, immunology and serology and internal medicine. The candidate in the third year is requested to present a thesis subsequently resuming his research work in the field of laboratory medicine throughout the final year under one of the teaching staff; the thesis is reviewed by an examining committee.

2. The certificate of specialist in one of the branches of laboratory medicine, of which clinical chemistry is the more important. The requirements for obtaining this certificate are three-years of training and full service in the laboratory of a university hospital, one of the main hospitals of the Ministry of Health, one of the two principal military hospitals or any other recognized foreign institution. After this training the candidate must pass an examination before a committee of examiners in the Ministry of Health. Medical doctors, pharmacists, chemical baccalaureates, graduates in natural sciences, veterinarians, etc., are eligible for this certification.

151

Place of clinical chemistry in hospital service

There are three university teaching hospitals in Syria, equipped with modern laboratories for clinical chemistry, immunology, microbiology, hematology and blood banking. The heads of these laboratories are medical doctors who have specialized in one of the branches of laboratory medicine, under whom the heads of the section of clinical chemistry may also be specialized pharmacists.

There are about 35 government-owned hospitals in Syria, all of which have laboratory facilities and these laboratories are mainly suited for routine testing. The most common investigations carried out are microscopy, routine blood counts and clinical chemistry and some blood grouping and crossmatching. Each government-owned hospital laboratory is directed by a 'specialist' with a staff comprising one senior technologist, an assistant laboratory technologist and two or three laboratory assistants.

The private hospitals are not equipped for laboratory analyses. They depend on private laboratories to do their analyses.

Place of clinical chemistry in non-hospital service

There are forty private medical laboratories. These laboratories cover the analytical service to the general practitioners as well as to the private hospitals.

The directors of these laboratories are either medical doctors or pharmacists, some of them holding appointments in some of the university teaching hospitals or the government-owned hospitals.

Place of clinical chemistry in medical education

The medical curriculum comprises 120 hours (lectures) of biochemistry in the second academic year, along with 60 practical hours annually, most of which deal with clinical chemistry. Students of the final year spend fifteen full successive days in the department of laboratory medicine attending seminars and practising different analyses, mainly: urinalysis, blood counts, blood banking, function tests, etc.

Technologists, technicians and other laboratory workers

Since 1972 laboratory technicians have been trained according to regulations issued by the Ministry of Health; this leads to certification as a registered laboratory technician.

The trainee must have his secondary school baccalaureate degree. The education takes a total of two years including:—

1. A six-months' period of practical training in an authorized department of laboratory medicine.

2. A 12-months' theoretical course at an authorized school for laboratory technicians (about 850 hours).

One school belongs to Aleppo University faculty of medicine, another to the Ministry of Health technical institute and a third one to the Palestinian Refugees UN Training Center (VTC). This theoretical course ends with an examination each year.

3. Practical training in their school for the same period as the theoretical course (about 1200 hours), ending with a practical exam.

4. Final practical training in the authorized department of laboratory medicine is ended by an examination.

Possible developments for the next ten years

Throughout the last four years, many steps have been taken to advance laboratory work, with special interest in automation and quality control.

It is anticipated that this will continue over the next decade. Ultra-micro methods will be adopted more and more. The number of individual constituents examined will increase. Tolerance tests as well as hormone immunoassays will be more common. Quality control procedures will be stressed. Additional more complex analyses will be added. The demand for training personnel will increase considerably.

National societies, associations and journals

Syrian Society for Clinical Chemistry
Founded in 1975. Number of members: 50. Concerned with the scientific and educational aspects of clinical chemistry.

Organization of Medical Analysis Laboratories
Under the Syrian Medical Association and the Syrian Pharmaceutical Association, founded in 1974. Number of members: 40. This organization takes care of the professional and economic interests of its members.

The Supreme Committee for Laboratory Control
This committee under the Ministry of Health was created to undertake the responsibility of:—

1. Supervising laboratory examinations in the different official and private laboratories in the country by a quality control program.

2. Specifying the minimal qualifications to be available in different laboratories as concerns space, apparatus and workers.

3. Standardization of methods used in different laboratories and alignment of the units utilized with international (SI) units.

The Arab Journal of Laboratory Medicine
Scientific journal edited by the Egyptian and the Syrian Societies for Clinical Chemistry and the Allied Arab Societies. Founded in 1975, appearing three times a year. Editorial Office, Department of Chemical Pathology, Kasser-el-Aini School of Medicine, Cairo University, Cairo, Egypt.

UNION OF SOVIET SOCIALIST REPUBLICS

History

At the beginning of the twentieth century, before the Great October Socialist Revolution, single chemical and biochemical investigations were carried out along with general chemical research in university clinics, chairs, and private hospitals. After the revolution, laboratories were formed parallel to the network of curative establishments.

Laboratories are established at different types of curative and prophylactic establishments (hospitals, polyclinics, dispensaries, consultation offices, medical and sanitary units at plants and factories). Originally laboratories were independent establishments with their own staff and finances, but after the twenties, laboratories were established at all more or less large medical establishments, and specialization took place in the main spheres of work: curative, sanitary and epidemiological, and pathoanatomic.

Clinicochemical research is an essential part of a complex of the general laboratory examination of a patient. At the same time a number of guides for clinicobiochemical investigations are issued (Balakhovsky, Petrunkina, Asatiana *et al.*].

Clinicochemical research at that time was intensively carried out at the Bach Institute of Biochemistry, the Academy of Medical Sciences of the USSR and in the biochemistry department of the Institute of Experimental Medicine of the Ministry of Health of the USSR which was later transformed into the Institute of Biological and Medical Chemistry of the Academy of Medical Sciences of the USSR.

Table 1

Rise in the number of clinical and diagnostic laboratories

YEAR	1913	1940	1955	1970	1972
No. of Laboratories	96	6,227	14,902	35,454	37,923

After the end of the Second World War (1941-1945), an increase in the number and equipment of laboratories proceeded at a great rate of expansion. Practical work is supported by numerous investigations carried out at the chairs of medical institutes and institutes of the further training of physicians as well as in research institutes.

Since that time clinical chemistry has developed in close contact with laboratory diagnostics. Subsequently we shall touch upon only the clinicobiochemical aspect of the work of clinical and diagnostic laboratories; however, one should bear in mind that laboratories are likewise closely connected with general clinical diagnostics except those situations which are mentioned separately. Organization of laboratory diagnostics, and of clinical chemistry in particular, is an important branch of work of the Ministry of Health of the USSR and Union Republics.

In 1968 the All-Union Scientific Center for Guidance on Methods for Laboratory Developments of the Ministry of Health of the USSR was formed. The Center united the organizational and methodic work in the sphere of laboratory diagnostics on the Union scale. In the same year the post of the chief specialist in laboratory diagnostics in the Ministry of Health of the USSR and the Ministries of Health of Union Republics was established. The specialist is responsible for the solution and co-ordination of questions concerned with the establishment of laboratory service. The system of guidance on methods for central laboratories and for laboratories of peripheral curative and prophylactic establishments was organized.

Standard methods for clinicobiochemical research have been developed, many clinicobiochemical analyses being provided by the standard sets of reagents.

Legal status

Health laboratory services which include clinicodiagnostic branches are guided in their activity first of all by general principles of Soviet public health. They are: free-of-charge universally accessible, and prophylactic medical care. These principles are embodied in the Foundations of the Legislation for Public Health in the USSR and Union Republics which came into force since 1972.

Concrete questions relative to the organization of laboratory work are regulated by general orders and instructions of the Ministry of Health of the USSR and Ministry of Union Republics. These instructions control hygienic requirements for working rooms, staff, the qualifications of individuals who have the right to work in diagnostic laboratories, the duration of the working day and annual vacation, the amount of wages in relation to qualification and length of service in the given field.

Education and advanced training of the heads of laboratories and personnel with higher education

There are the following categories of specialists with higher education in clinical chemistry:

(a) Physicians.
(b) Pharmacists.
(c) Biologists who graduated from the university or a pedagogical institute.

(d)   Chemists who graduated from the university or a specialized higher educational establishment*.
(Primary specialization and advanced training are mentioned subsequently).

A laboratory in a clinical establishment of a curative and prophylactic type is headed by a physician who has sufficient experience in this field and was approved in the first or higher category, whereas persons with non-medical education (biological or chemical) may not occupy leading posts and can work only as specialists. In research institutes clinico-chemical laboratories are headed by persons who have the doctor's or bachelor's degree in medical or biological sciences.

For the training of personnel in clinical chemistry for research establishments, use is made of postgraduate training and other channels. The law in the Soviet Union provides for the competition for the candidates degree and the degree of the doctor of science. The degree is accredited as a result of a public defense of a specially carried-out research or a series of papers published in scientific literature. A widespread way of training competent personnel of clinical chemists is 'aspirantura' — postgraduate training (a three-year course, as a rule, with the public defense of a candidate's thesis at the end) and 'ordinatura' (two-year course) in the chairs and laboratories where there are favourable conditions and necessary guidance. Competitors for the candidate's degree must pass preliminarily, examinations including those of the special subject.

Place of clinical chemistry in hospital examination

In the USSR, as well as in other countries, clinicochemical examination of patients and the conduct of prophylactic analyses may be exercized by a clinicochemical laboratory, headed by the corresponding chief. However, it is more common practice to do clinicochemical work in a branch of a general clinicodiagnostic laboratory.

It is worthy of note that there is a clear cut differentiation of laboratory specialities in Soviet medicine. Clinical chemists deal directly with the investigation of the ingredients of the organism of the patient; chemists working in the system of sanitary and epidemiological stations are concerned with the study of pathogenic factors of the environment; chemists who work in pathoanatomical laboratories examine cadaver material.

Clinicodiagnostic laboratories in the USSR are divided into five categories depending on the number of hospital beds and the number of physicians who receive patients in the polyclinic. Lists of obligatory investigations have been compiled for the laboratory of the given category. Clinicochemical investigations (depending on the category) constitute from 28% to 40% of the total number of different types of

---

* In recent years, special departments have been formed at several universities for the training of specialists in clinical chemistry.

investigations. The rest of the investigations are represented by hemato-logical, cytological, microbiological and immunological analyses. It should be pointed out that the annual rise in the number of analyses made in laboratories is mainly due to the clinicochemical investigations.

Along with clinicodiagnostic laboratories of the general type, specia-lized laboratories are established, among them: enzymological, hor-monal, coagulation, chemicotoxicological, as well as laboratories of clinicochemical 'stat' diagnostics.

The formation of centers and departments of reanimation and chronic hemodialysis led to the formation of specialized clinicochemical labora-tories in these departments.

Place of clinical chemistry in non-hospital services
Laboratories of polyclinics are given equivalent status in power and equipment. Laboratory investigations are an essential part of mass prophylactic examinations of the population. Various programs of laboratory screening are exercized (revealing hereditary diseases, onco-logical diseases, atherosclerosis etc.), as well as dispensary screening and permanent observation of certain groups of population (groups at risk).

Place of clinical chemistry in medical education—clinical chemistry in the program of medical institutes
In the general course of medical institutes training in clinical chemistry occupies a considerable place. In the first year of study students enlarge their knowledge in the field of organic and physicocolloid chemistry obtained at secondary school. Later on they are given a course of general and medical biochemistry. Examinations are to be taken in these subjects; the study of these subjects takes three semesters in the general six-year program of training of physicians. Further, while studying the course at the corresponding chairs, medical students are given courses of laboratory diagnostics of special groups of diseases. In recent years, a faculty has been formed at one of the Moscow medical institutes to train specialists in laboratory work. Persons graduating from this faculty gain corresponding diplomas.

Clinical chemistry and specialization after graduating from the institute
There are several ways of training specialists in clinical chemistry after graduating from the medical institute.

Those physicians who are eager to devote themselves to laboratory diagnostics, upon graduating from the institute join courses of speciali-zation at numerous institutes of further training of physicians or specia-lize directly at their working place in laboratories.

As mentioned previously, guidance is given on the part of laboratories of central curative and prophylactic establishments for the peripheral laboratories. Such guidance provides for training of specialists for peripheral regions at their working places in laboratories of central establishments.

The second channel for training competent staff is that of specialization in the field of clinical chemistry in the chairs of clinical chemistry and laboratory diagnostics at the institutes for advanced medical training.

Further training of laboratory workers with higher education who have certain practical experience in this field, takes place in the same chairs. Such advanced training is essential for gaining a higher category*. Persons who are eager to join courses are obliged to perform certain assignments and present an abstract. Those who finish courses of specialization or advanced training are given appropriate certificates.

In the program of specialization of non-laboratory physicians (pediatricians, therapeutists, surgeons *et al.*), training in clinical chemistry occupies a certain place also.

Place of clinical chemistry in non-medical education

For persons who have no medical training and work in clinicodiagnostic laboratories, special courses are organized at the institutes of advanced medical training. Several subjects concerned with human pathology are added to the programs of these courses.

As previously mentioned, persons who are going to write a thesis in the field of clinical chemistry, while in preparation for the competition of a degree, are required to take, in particular, an examination in clinical chemistry.

Laboratories, technicians and other laboratory workers

Persons with secondary special education work in the field of clinical chemistry as medium-grade personnel†. They are trained at schools of medical technology and upon graduation receive corresponding diplomas. Some persons who occupy the posts of laboratorians in curative and prophylactic establishments, after finishing schools of general education are trained at their working places. In this case, however, there are differences in payment between those who have a diploma in special training and those who have none.

Sanitarians (persons who are in charge of the tidying of premises, washing of plates and dishes, etc.) and technical assistants (persons who are allowed to carry out separate simple operations such as distilla-

---

* In brief, the education system in the USSR is represented by the following main stages: (1) school of general education, (2) technical schools which give secondary special training to persons who finished eight or ten classes of secondary school, and (3) institutes and universities which graduate persons with higher education.

† In the USSR, depending on the level of training and length of service, physicians are divided into three categories: the second, the first and the highest. The category is assigned by the resolution of a special commission of local organs of public health, which interviews a candidate for a higher category and considers the results of his work.

tion of water separation of plasma from erthrocytes with the help of centrifuging, the measurement of the total amount of biological fluids sent for analysis, etc.) are occupied by persons who have secondary or elementary education; special training in this course is not obligatory.

It should be mentioned that recently a significant part in clinico-biochemical laboratories has been played by engineers who provide for the maintenance of the numerous apparatus used in these laboratories.

Possible future developments
Analysis of data for recent years indicates that there is a steady relative and absolute rise in the number of clinicochemical analyses which in its turn raises certain serious questions in the sphere of planning of funds, resources, personnel training and scientific organization of labour. At the same time it is necessary to note that increase of diagnostic value of such investigations undoubtedly justifies the expenses.

Public organizations and journals
In the USSR clinical chemists are mainly members of the two societies. These are; the All-Union Biochemical Society and the All-Union Society of Laboratory Diagnostics.

The first society unites biochemists of different profiles, clinical chemistry being represented by a special section. The second one unites physicians of different background working in the field of laboratory diagnosis of diseases. This is one of the largest medical societies, a considerable number of whose members work in the field of clinical chemistry. It should be also mentioned that a certain per cent of clinical chemists are members of the All-Union Society of Pathoanatomists— these are mainly specialists occupied with bordering regions of clinical chemistry and histochemistry, for example, specialists concerned with biochemistry of malignant tumours. Societies of laboratory diagnostics and pathoanatomists are accredited at the Ministry of Health of the USSR. Upon the approval of the Ministry of Health and its financial support these societies hold their meetings (annually, as a rule) and congresses (once in four years) at which various problems of clinical chemistry are discussed (raised either as main questions or at least as a part). Financial support is provided by the Ministry of Higher Education of the USSR and the USSR Academy of Sciences.

As stated above, early in 1930, the first guides for clinical chemistry were issued. Later on their number increased at an exponential rate; at present there are many general guides and handbooks as well as handbooks for special branches (study of hormones, enzymes, proteins). Since 1925 the publication of the journal *Laboratornoyedelo* (formerly called *Laboratornaya Praktika*) commenced; this is one of the oldest medical publications in the Soviet period; the journal deals with methodological and organizational aspects of clinical chemistry.

159

In 1954 the journal *Voprosyi meditsinskoikhimii* was initiated. There are permanent special sections as well as single articles in clinical journals, devoted to the problems of clinical biochemistry of various diseases.

In working out recommendations and instructions concerning different aspects of laboratory service, the Ministry of Health of the USSR relies upon several public councils and commissions, including specialists in clinical chemistry, both scientists and practitioners.

A list of commissions:

(1) Consultative Council for Laboratory Developments.

(2) Commission for Organization and Guidance on Methods.

(3) Commission for Unification and Control of Quality of Clinical Laboratory Methods of Research.

(4) Commission for Laboratory Equipment.

(5) Commission for Reagents.

(6) Commission for Unification of Methods of Research of the System of Blood Coagulation and Fibrinolysis.

Bibliography

Balakhovsky, S. D., Balakovsky, I. S., Metody khimicheskogo analiza krovi. Izd. 3-e M., Medgiz 1953, 754x.x il. Il. skhemy.

Balakhovsky, S. D., Mikrokhimicheskyi analiz krovi i yego klinicheskoye znacheniye. M.-L., Gosizdat, 1969, 652s.

Bilkhimicheskiye metody issledovaniya v klinike. (Spravochnik). Pod. red. A. A. Pokrovskogo. M., *'Meditsina'*, 1969. 652s.

Vvedeniye v klinicheskuyu biokhimiyu. (Osnovy patobiokhimii). Pod. red. I. I. Ivanova. L., *'Meditsina'*, Lenigr. otd.-niye, 1969, 493s. s.il.

Zbarsky, B. I., Ivanov, I. I., Mardashev, S. P., Biologicheskaya khimiya. (Uchebnik dlya med. in-tov). Izd. 5-ye, ispr. i don. L., *'Meditsina'*, Lenigr. otd.-niye, 1972, 582s.s.il., Il.il.

Kurs klinicheskoyi biokhimii. Ucheb. posobiye k zanyatiyam po klinich. biokhimii. Pod. red. V. V. Menshikova, Vyp. 1,2,3. M., 1969. (I. Moskovskyi med. int).

Larsky, E. G., Rubin, V. I., and Solun, N. S., Biokhimicheskiye metody issledovaniya v klinike. Metod. posobiye dlya vrachej-la-borantov i slushatelej phak. usovershenstvovaniya vrachej. Saratov, 1968. 309s. (Saratov. med. in-t. Saratov. obl. otd. zdravookhaneiya).

Larsky, E. G., Metody zonalnogo elektrophoreza. M., *'Meditsina'*, 1971, 112s.s.il.

Menshikov, V. V., Izbrannye lektsii po klinicheskoj biokhimii. Ch.1,2, M., 1970. (I. Moskovskyi med. in-t).

Menshikov, V. V., Metody klinicheskoj biokhimii gormonov i media-torov. (Ucheb. posobiye). Izd. 2-ye pererab. i dop. M., 1969 159s. (I. Moskovsky med. in-t, Kurs Klinicheskoj biokhimii).

Nauchnyye meditsinskiye obtshestva SSR. Sb. Statej. Pod red. M. V.

Volkova. M., *'Meditsina'*, 1972. 456s. (Uchen.med.sovet.Sovet Nauch. med. o-va).

Petrovsky B. V. Zdorovye naroda-vazhnejsheye dostoyaniye sotsialisticheskogo obtshestva. M., *'Meditsina'*, 1971. 104s.s.il.

Petrunkin, M. A., Petrunkina, A. M., Prakticheskaya biokhimiya. Izd. 2-ye, ispr.i.dop. L., Medgiz, Leningr.otd.-niye, 1951. 359s.

Petrunkina, A. M., Prakticheskaya biokhimiya. Izd. 3-ye, pererab. L., Medgiz, Lenigr. otd.-niye, 1961. 427s.s.ill.

Rukovodstro po klinicheskim laboratornym issledovaniyam, osnovannoye V. E. Predtechenskym. Pod red. E. A. Kost i L. G. Smirnovoj. Izd. 6-ye, stereotip., M., *'Meditsina'*, 1964, 960s.s.il., 291., il.

Savchenko, M. G., Kratky ocherk isotorii razvitiya laboratornoj klinicheskoj diagnostiki. Tashkent, Medgiz UzSSR, 1960. 60s.s.il.

Spravovhnik po klinicheskim laboratornym metodam issledovaniya. Pod. obtsh. red. E. A. Kost. M., *'Meditsina'*, 1968, 436s. s.il., 4 l., il.

Syagayev, S. A., Menshikov, V. V., Makarova, N. A., Razvitiye laboratornoj sluzhby za 50 let sutshestvovaniya Soyuza SSR. *'Labor.delo'*, 1972, 12, 707-714.

Isakov, U. F. The System of medical education in the USSR. World Health Organization Inter-Regional Travelling Semihar on the Organization and Functioning of Hospital Sanitary and Epidemiological Laboratory services, 3-26 April, Moscow, 1968. 21p. (Central Institute for Advanced Medical Studies).

Tatarinova, S. D. The Training, Specialization and Advanced Training of Laboratory workers of Soviet Public Health Services, World Health Organization Inter-Regional Travelling Seminar on the Organization and Functioning of Hospital Sanitary and Epidemiological Laboratory Services. 3-26 April Moscow, 1968. 21p. (Central Institute for Medical Studies).

UNITED KINGDOM

Development of clinical chemistry in the United Kingdom

The development of clinical chemistry in the United Kingdom goes back to the seventeenth century, during which period Thomas Willis wrote a dissertation on urine and described the presence or absence of a sweet taste in relation to diabetes. In this period also, Robert Boyle made observations on blood coagulation and listed a number of headings for urine examination. During the nineteenth century, Wollaston, a physician, described the variety of chemically distinct types of urinary calculi and in 1836 Richard Bright associated the presence of albumin in the urine with kidney disease. Later, Henry Bence Jones had his name associated with the 'protein' which appears in the urine of some cases of myelomatosis. At this period, appointments were beginning to be made in medical schools and J. W. L. Thudichum became lecturer in chemical pathology at St. Thomas' Hospital, London. It was also during the nineteenth century that Sir Alfred Garrod devised one of the first quantitative methods applied to blood. This was the estimation of uric acid, which was made to crystallize on threads of a material called huckaback. In 1838 Rees demonstrated the presence of glucose in diabetic blood.

During the twentieth century, and particularly in the period between the two world wars, clinical chemistry developed rapidly and tended to be known as chemical pathology, especially in the South of England. University professional appointments concerned with the subject were made. Charles Dodds (later Sir Charles) took up the chair of biochemistry at the Middlesex Hospital Medical School in 1927. He was responsible also for the clinical chemistry laboratory of the hospital. In 1934, Earl J. King was appointed reader in chemical pathology and head of that department at the Postgraduate Medical School. In 1944 he was appointed reader in chemical pathology in the University of London, his appointment being held at The Royal Postgraduate Medical School, St. Bartholomew's Hospital.

In 1946, the University of Edinburgh established a department of clinical chemistry with C. P. Stewart as its head. During the postwar period, a number of chairs were created throughout the United Kingdom and such appointments now exist at most of the medical schools of the University of London, as well as in those of the Universities of Edinburgh, Glasgow, Aberdeen, Dundee, Leeds, Newcastle upon Tyne, Birmingham, Cardiff, Surrey, Cambridge, Oxford and Southampton. In London, the title chemical pathology is used, whereas in the rest of the country the chairs are usually titled either clinical chemistry or clinical biochemistry. In addition to the senior academic appointments, non-teaching hospitals of appropriate size also have clinical chemistry laboratories appropriately staffed. At first it was the rule for the discipline to be part of the general pathology laboratory, it is now the tendency in at least some of the larger hospitals for clinical chemistry to be autonomous. It must of course be emphasized that there have been clinical chemistry appointments at hospitals for the last 50 years or so. These were frequently, but not invariably, held by physicians, with technical, and later science graduate, assistance.

The standard practical textbooks of this early period were *Chemical Methods in Clinical Medicine* by G. A. Harrison (1st edition, 1930), *Practical Clinical Biochemistry* by H. Varley (1st edition, 1954) and *Microanalysis in Medical Biochemistry* by E. J. King (1946).

As time progressed, the workload in the laboratories, as elsewhere in the world, increased enormously, averaging some 17% per annum. In the United Kingdom, as elsewhere, it was only possible to carry these loads by introduction of automatic procedures. In addition to this workload increase, there was also an enlargement of the spectrum of different types of analyses performed and the introduction of specific procedures such as immunochemistry and radioimmunoassay. It soon became obvious that for certain procedures special expertise was required. It was difficult to achieve such skills from the limited number of assays available on a local basis and this tended to lead to the use of research laboratories for such techniques as special steroid analyses. The Department of Health and Social Security, therefore, decided to set up a number of supraregional assay centres, which undertake special analyses for the whole of the country. These supraregional laboratories cover the fields of peptide hormone assays, determinations by radio-immunoassay, special steroid analyses, specific protein analyses by immunochemical methods (for example carcinoembryonic antigen and a-fetoprotein, as well as the immunoglobulins). The determination of lead can also be obtained as a supraregional service. It must be pointed out that if local laboratories have appropriate equipment and expertise, they are not necessarily required to use the supraregional assay service.

Education and training for director and senior staff of hospital clinical chemistry laboratories—medical graduates

The attainment of senior status in chemical pathology for a medical graduate follows a definitive educational sequence which is marked by stepwise advancement.

Grading of appointments

[a] *Preregistration period*—Medical graduates must hold preregistration appointments for six months in medicine and six months in surgery in recognized hospitals before they can be legally registered.

[b] *Senior house officer*—After registration, a further period of one year is spent in appointment at this grade. For those ultimately specializing in clinical chemistry, this is usually in general pathology.

[c] *Registrar*—For an additional two years appointments are held at this grading and most of the time is spent in chemical pathology. It is usual at this level to take part 1 of the examination for membership of the Royal College of Pathologists (MRCPath), or the MSc in clinical biochemistry.

[d] *Senior registrar*—This appointment is usually held for from three to five years, and to enter this grade the candidate is usually expected to hold some sort of higher qualification. In pathology this would normally, although not invariably, be MRCPath. (pt. 1).

[e] *Consultant grade*—At this level the candidate has full consultant responsibility. He usually has at least eight years' postgraduate experience and must have the necessary higher qualifications. Appropriate publications are of value in obtaining an appointment.

The training of medical graduates is now very much under the guidance of the Royal College of Pathologists, which insists that candidates may not sit its examinations unless training has taken place for appropriate periods in laboratories recognized and inspected by the college. The college awards a membership of the Royal College of Pathologists (MRCPath) which can be taken in the various branches of pathology. A scheme of study has been prepared by the college.

A chemical pathologist in training spends a preliminary (general) training period extending over approximately three years. It is recommended that experience should be gained in all four major branches of pathology during this period, with a minimum of six months in each of the other departments (hematology, microbiology and morbid anatomy) and at least one year in a chemical pathology department. In certain cases these requirements may be waived.

At the end of the preliminary training period taken during his senior house officer and registrar appointment, the candidate sits the primary examination of the Royal College of Pathologists and, if successful, then proceeds to the advanced (specialist) training period. The primary examination of the college is now taken in one subject only, and consists of written practical and oral examinations.

The advanced (specialist) training period is usually undertaken while holding a senior registrar appointment and after the trainee has passed the primary examination of the college.

The advanced training period extends over a minimum of three years and towards the end of this the trainee is expected to pass the final examination and become a Member of the Royal College of Pathologists. The trainee is expected to extend his experience of the techniques and procedures to which he was introduced in his earlier training. In addition, he receives tuition in the theory and practice of chromatography, electrophoresis, spectroscopy, immunoelectrophoresis and the use of radioactive isotopes in diagnostic procedures. He learns to handle automatic analyzing equipment including control procedures and is introduced to the theory and practice of data handling and the specific methods which this involves. Most of this training is given under the supervision of senior members of the laboratory staff. Trainees can also attend special courses arranged by the Association of Clinical Biochemists.

Expenses for attending the courses or the meetings may be met by the National Health Service.

Arrangements are made either by day or block release for trainees to continue basic scientific training designed to complement previous basic training. It is important that trainees should acquire a sound basis of general chemistry and biochemistry and should gain a working knowledge of chemical genetics and statistics.

The department at Newcastle upon Tyne has unique facilities whereby medical graduates during their registrar and early senior registrar appointments, can take full university courses in chemistry leading to the degree of BSc (Honours) in chemistry.

At the Universities of Birmingham, London, Newcastle upon Tyne and Surrey courses have been established leading to the degree of MSc in clinical biochemistry. Such a degree may exempt from part 1 of the MRCPath.

Medical graduates can also sit for the Mastership in clinical biochemistry.

It should be pointed out that the qualification of Membership of the Royal College of Pathologists can also be taken by non-medical graduates. If such individuals already hold the Mastership in Clinical Biochemistry, they are exempted from the necessity of sitting the primary examination. Opportunities also exist in certain cases for candidates to be elected to the Membership on the basis of submission of research publications.

Education and training for non-medical graduates

Qualification and grading

The qualification necessary is an appropriate science degree of a university of England, Wales, Scotland or Ireland or Membership of the Royal Institute of Chemistry or such other qualification.

The various grades are as follows:

[a]   *Basic Grade*—A biochemist who is qualified as set out above and is employed under appropriate professional supervision. There is a probationary period of at least two years for all new entrants to this grade. Retention in the service is then conditional upon the acceptance of a certificate of proficiency by the employing authority.

[b]   *Senior Grade*—A biochemist who is employed in a post of greater responsibility, the duties of which might or might not include the supervision of biochemists in the basic grade and who has had at least four years' experience in the Basic Grade.

[c]   *Principal Grade*—A biochemist who is in charge of an important section of a large biochemistry department or who is in charge of the whole of a smaller department or who has shown exceptional merit or is doing work of a special nature.

[d]   *Top Grade*—A biochemist in charge of a large department. The application of this grading is subject to the concurrence of the Minister of Health or the Secretary of State for Scotland.

Academic posts—Professional chairs, variously designated chemical pathology or clinical biochemistry, have been created at several universities in the UK. The academic departments attached to these chairs have the usual academic posts of reader, senior lecturers and lecturers, and these departments are normally situated in university hospitals and often work in close association with the service department of the hospital. The professor is normally head of both academic and service departments, although there are local variations. Such departments offer an alternative career structure to the NHS for both medically and scientifically-qualified graduates. Naturally, academic posts will best suit those with an inclination towards teaching or research, and research degrees such as the MD or PhD may be given more emphasis for appointment to the various grades than Membership of a Royal College or the MCB.

Educational facilities
The Association of Clinical Biochemists arranges special training courses at various centers. Expenses for such courses are usually met by the National Health Service but permission for absence must first be obtained from the employing authority.

A good deal of education is of the apprentice type whereby trainees learn at the laboratory bench under the supervision of a senior member of the staff. They also attend local and national meetings of the Association of Clinical Biochemists and occasionally special courses under the aegis of the Association of Clinical Pathologists.

Qualifications
The available qualifications are a Mastership in science and a Mastership in clinical biochemistry.

[a] *Mastership in Science*—Courses leading to the MSc in clinical biochemistry are now held in five universities. The University of Surrey, Leeds and the University of Birmingham, London and Dublin run two-year courses. In Surrey and Leeds they are attended by individuals already in post in the Basic Grade (usually probationary).

In Birmingham students are in supernumerary posts and are paid by the Regional Health Authority.

The University of Newcastle upon Tyne runs a full-time course for individuals who have just graduated.

The University of Leeds has also started an MSc course in steroid endocrinology.

In most universities it is possible also to take an MPhil by research work in clinical chemistry. In laboratories of certain teaching hospitals associated with a university it is possible to undertake research and present a thesis for the PhD.

[b] *Mastership in Clinical Biochemistry* [MCB]—This is a qualification awarded jointly by the Royal Institute of Chemistry, The Royal College of Physicians, The Royal College of Pathologists and the Association of Clinical Biochemists.

Very great importance is placed on suitable training and the Joint

Examinations Board requires evidence that a candidate has had a minimum of three years' practical postgraduate training in a post in clinical biochemistry before admission to the examination and a total minimum period of five years of appropriate experience is required as a condition of award.

No formal course of training is specified but during the whole of the five-year period the candidate is required to have been practising his profession under the general supervision of a principal or top grade biochemist in the National Health Service or under a consultant in chemical pathology or the holder of a university post of equivalent status. While there are obvious advantages in the closest possible working association between the candidate and his supervisor, it is not essential that they should be in the same laboratory, or even in the same hospital, provided they are in associated hospitals in the same Region. In this case there should be weekly consultations and regular second-ment of the candidate for a few weeks at the supervisor's hospital to ensure adequate direct supervision.

The formal examination comprises three three-hour papers in general biochemistry, clinical chemistry and analytical chemistry together with a practical examination of several days' duration. A further oral exami-nation is taken when the candidate has completed the five years of appropriate experience.

To take this qualification a candidate has his practical training largely while in post, does a good deal of private study and, if possible, attends the appropriate courses arranged by the Association of Clinical Bio-chemists. The Association Education Committee has also drawn up a scheme of study; in fact there is no official syllabus put out by the Examination Board. The Association of Clinical Biochemists also recommends to its members appropriate textbooks and the like.

Under the latest regulations, the formal examination at the end of three years is now known as Part 1 of the MCB and candidates who pass both the theoretical and practical sections of this part of the examination are awarded the diploma in clinical biochemistry. The MCB itself is still awarded after five years, but at this stage the examination has come to be known as Part 2. The examination at this stage is normally directed towards the assessment of the candidate's maturity and competence, his knowledge of laboratory management and, in particular, his ability to take independent charge of a hospital clinical biochemistry labora-tory.

Candidates are required to undergo an oral examination, which may be general or specialized according to the candidate's experience. In making their final assessment, the examiners are assisted by evidence indicating a contribution to laboratory medicine. This may consist of:

(1) published papers.

(2) a dissertation on some aspect(s) of laboratory work or develop-ment, or on the role of the biochemistry laboratory in the hospital service.

(3)   a report on a study in depth of some aspect of clinical bio-chemistry or biochemistry in relation to medicine.

(4)   a successfully submitted thesis for a higher degree.

Education and training for technicians

Qualifications and promotion grades

Minimum qualifications for entry is a pass in four subjects at the Ordi-nary Level of the Certificate of Education. These subjects must be English language, mathematics and two science subjects.

The various grades are as follows:

[*a*]   *Junior technician A*—This is a person who is training to become a state-registered technician and takes part in the technical work of the laboratory.

[*b*]   *Junior technician B*—This is a person who is training to become a state-registered technician and takes part in the technical work of the laboratory and has passed the Ordinary National Certificate or Ordinary National Diploma Examination in medical laboratory sciences.

[*c*]   *Medical laboratory technician*—This is a state-registered tech-nician who has passed the HNC or HND in medical laboratory sciences.

[*d*]   *Senior technician*—This is an experienced state-registered techni-cian who holds the Fellowship of the Institute of Medical Laboratory Sciences (or equivalent qualification) who is

(i)   in technical charge of a laboratory which employs, including the senior technician, not less than four technicians, or
(ii)   in technical charge of a section or department in which there are employed, including the senior technician, not less than four technicians, or
(iii)   regularly and mainly engaged on individual work requiring special skills and responsibility.

[*e*]   *Chief technician*—This is an experienced state-registered techni-cian who is:
(i)   in technical charge of a laboratory which employs not less than ten technicians including the chief technician, or
(ii)   in technical charge of a department of a laboratory where the combined departments employ not less than 25 technicians, or
(iii)   in technical charge of a section of a department in a laboratory where the combined departments employ not less than 63 tech-nicians providing that he is responsible for seven or more technicians, or
(iv)   wholly engaged on individual and important work involving complex examinations or research of a highly skilled nature.

[*f*] *Senior chief technician*—This is an experienced state-registered technician who:

(i) has overall technical charge of a laboratory where the combined departments employ, including the senior chief technician, not less than 25 technicians, or

(ii) is in technical charge of a department of a laboratory where the combined departments employ not less than 63 technicians, or

(iii) is in technical charge of a specialized laboratory organized independently of the main department and undertaking complex work not covered by the main departments.

[*g*] *Principal technician*—This is an experienced state-registered technician who has overall technical charge of the combined departments of a laboratory in which are employed, including the principal, not less than 63 technicians.

Educational facilities

Much of the teaching is done by the senior technicians in the actual laboratory in which the individuals are working and many laboratories run classes, some held during working hours and some in the evening. All technicians also attend classes at colleges of further education and technical colleges. This means they are away for one day a week and, in addition, attend an evening class each week. The junior technicians attend classes which lead to the examination known as the Ordinary National Certificate (ONC) in medical laboratory sciences. This makes him a junior technician, B. He then attends classes for a further two years which train him for the Higher National Certificate (HNC) in medical laboratory sciences with additional specialist training in biochemistry.

In addition to the Higher National Certificate it is also possible to take a Higher National Diploma which is rated as of equal significance. The difference is that whereas the Higher National Certificate is obtained through the system of day release already described, the Higher National Diploma is obtained by periods of release for up to three months per annum taken en bloc.

State registration and technician qualifications

To become a state-registered medical laboratory technician it is now necessary to hold a Higher National Certificate or Higher National Diploma in medical laboratory sciences. In addition the candidate must have spent a stated number of years working in a recognized laboratory. There is an opening for science graduates to become state-registered technicians after working for one year following graduation in a recognized laboratory.

The Institute of Medical Laboratory Sciences has, in the past, awarded an Associateship (AIMLS) on the result of HNC or HND Examination and a Fellowship based on the result of an examination set by the IMLS on submission of a research thesis.

The Association of Clinical Biochemists

By the early 1950s, it became apparent that clinical chemistry in the United Kingdom had reached a stage where it ought to have a separate representation of its own rather than simply a loose association with the Biochemical Society. In the Midlands, and the southern part of England a group of individuals gathered together in the early 1950s and very rapidly incorporated others from the rest of the country. A working party was set up to produce an appropriate constitution for the proposed new society. This was drawn up largely by E. J. King and A. L. Latner and finalized after a number of long discussion meetings. It has, of course, been very much modified. In 1953, came the actual formation of the Association of Clinical Biochemists. The first president was N. F. Maclagan (1953-1955 inclusive), he was followed by E. J. King (1956-1958 inclusive), C. P. Stewart (1959 and 1960), A. L. Latner (1961 and 1962), J. N. Davidson (1963 and 1964), R. H. S. Thompson (1965 and 1966), R. Gaddie (1967 and 1968), C. H. Gray (1969 and 1970), A. Neuberger (1971 and 1972), and J. H. Wilkinson (1973 to 1975 inclusive). The current president is I. D. P. Wootton. The first chairman was E. J. King (1953 and 1954), followed by C. P. Stewart (1955 and 1956), R. Gaddie (1957 and 1958), A. L. Latner (1959 and 1960), H. Varley (1961 to 1963 inclusive), R. Gaddie (1964 to 1966 inclusive), G. H. Lathe (1967 to 1969 inclusive), F. L. Mitchell (1970 to 1972 inclusive), P. D. Griffiths (1973 to 1976) and the present chairman is A. M. Bold. The first treasurer was I. D. P. Wootton and the first secretary, A. L. Tarnoky. In 1963, the association appointed a meetings secretary. The first holder of this appointment was D. W. Moss. In 1973 an assistant secretary was also appointed and the first holder of this appointment is Miss P. R. N. Kind.

It was originally agreed that the country be divided into five regions, namely (i) the South-east and South-west, (ii) the North-west, (iii) Scotland and Northern Ireland, (iv) the Midlands, and (v) the North-east. At the beginning, the council of the association consisted of a member elected from each of these five regions and two national members. There are now three national members and six regional members, since there have been six regions as from 1971. Instead of the single South-east and South-west region, there is now the Southern England region, and the South Wales and South-west region.

In 1961, council proposed the formation of a professional committee, which was approved by the association. The first chairman of this committee was S. C. Frazer, and the first secretary, T. P. Whitehead. In 1963, the scientific committee was formed, its first chairman was A. L. Latner and its first secretary, A. H. Gowenlock. This committee has now become the scientific and technical committee. In 1970, the education committee was formed, the first chairman being H. G. Sammons and the first secretary, A. M. Bold. The publications committee was established in 1973. The first chairman was D. N. Raine, and

170

secretary was D. H. Orrell. Of course, the association had been publishing its *Proceedings* long before this, the first number appearing in August 1960, under the honorary editorship of H. Varley. Prior to the *Proceedings,* the association published a *Newsletter,* which still continues.

Professional qualifications

In 1962, A. L. Latner, who was then president of the association, commenced negotiations with Sir Charles Dodds, president of the Royal College of Physicians, with a view to getting closer co-operation between physicians and clinical biochemists. It was eventually decided that this could, in the first place, best be done by arranging that the Royal College of Physicians had some interest in the qualifications of clinical chemists. The Royal Institute of Chemistry was also involved at this stage. Latner was succeeded by J. N. Davidson, who took the matter further, and eventually, a joint examination board was formed consisting of representatives of the Association of Clinical Biochemists, the Royal College of Pathologists, the Royal College of Physicians and the Royal Institute of Chemistry. This board set up appropriate regulations for the diploma of Mastership in clinical biochemistry (MCB) and very shortly afterwards examinations were instigated. This qualification is now regarded as one of the higher qualifications required for senior biochemists, especially laboratory directors. One cannot speak too highly of the efforts of I. D. P. Wootton in the early days of the MCB, in relation to arranging the appropriate practical examinations, as well as his work on the examinations board itself. The first chairman of the board was Sir Charles Dodds. The Royal College of Pathologists recognize the MCB as one of the means of exemption from the primary examination for the Membership of the Royal College of Pathologists (MRCPath) and a number of clinical chemists, medical and non-medical, now hold this latter qualification. From the foregoing, it is obvious that close relationship has been maintained with the Royal College of Pathologists from the early days of formation of that body.

Quality control

A quality control scheme involving voluntary participation by biochemistry laboratories run by members in the United Kingdom started in July, 1969. The scheme is operated from the department of biochemistry, Queen Elizabeth Hospital, Birmingham, under the auspices of the standardization sub-committee of the laboratory equipment and methods advisory group of the Department of Health and Social Security. The findings are confidential. In the United Kingdom there are over 400 hospital biochemistry laboratories most of which participate in the scheme. In the near future. all biochemistry laboratories are to be circulated and invited to join in the scheme.

It is a matter of some interest that in the early 1950s a joint working party of the Association of Clinical Biochemists and the Association of Clinical Pathologists met under the able guidance and leadership of E. J. King. This working party tackled a number of important problems of mutual interest to the two associations. One of these was the question of preparation of standard sera. Such standards were originally prepared by Wellcome and I. D. P. Wootton played a most important part in maintaining a relationship with the firm and in distribution of the sera they prepared. At first the sera were checked by six of the country's leading laboratories and subsequently distributed to as many laboratories throughout the country as were prepared to undertake analyses. I. D. P. Wootton studied their results on a statistical basis and showed that there was an enormous range of variation in the country as a whole from laboratory to laboratory. This led to Wootton's very important investigations which demonstrated gross inaccuracies in many laboratories, not only in the United Kingdom but in countries throughout the world, including the USA. These findings were the initial impetus for the whole concept of quality control.

Association with other bodies
The Association of Clinical Biochemists has good relationships with both the Biochemical Society and the Association of Clinical Pathologists. In fact it holds joint meetings with each of these bodies. There is, of course, close association with the Royal College of Pathologists and a member of the Association of Clinical Biochemists is nominated to the advisory committee in chemical pathology of the college. The Royal College of Physicians is, of course, involved in relation to the MCB examination. The Royal Society has arranged that the Association of Clinical Biochemists nominates one member to the British National Committee for biochemistry and another member to the British National Committee for chemistry. The first member sitting on each of these two committees were C. P. Stewart and A. L. Latner respectively. This association with the Royal Society is a highly valued one and does much to indicate the fact that scientists of worth recognize the scientific stature of the clinical biochemist. There is, of course, very close association with the Department of Health and Social Security.

Publications
It has already been mentioned that the society publishes a *Newsletter* as well as the *Annals of Clinical Biochemistry,* the latter now being published in association with the British Medical Association, under the general editorship of N. F. Maclagan. Since 1974, the association has also been publishing *Current Clinical Chemistry,* which is a monthly publication giving current literature citations in the form of references to a very wide range of published papers dealing with clinical chemistry. These references are appropriately subdivided into an appropriate number of sections dealing with different aspects of the subject. *Current*

Clinical Chemistry also contains a list of new books of interest to the clinical chemist. This publication has become very greatly appreciated and is regarded by members of the association as a very significant advance. The association is indeed greatly indebted to J. G. Lines, who is the editor. He is, of course, assisted by a group of editorial advisors. In collaboration with the Biochemical Society, the association is currently producing *'Essays in Medical Biochemistry'*, under the joint editorship of V. Marks and C. N. Hales. This is a publication similar to *Essays in Biochemistry*, published by the Biochemical Society.

Possible future developments
It seems very likely that the association will separate its professional activities from its scientific activities. There is a feeling that the former is becoming highly complicated and is a type of union activity, which should be carried on by a separate body, which would fit in with the general regulations for unions in the United Kingdom. The association itself would then concentrate on scientific matters.

Many think it desirable for a consultant chemical pathologist to have considerable postgraduate clinical experience, and perhaps even the Membership of the Royal College of Physicians. However, the extent to which the chemical pathologist with medical training will emphasize the clinical as opposed to the laboratory side of his responsibilities still depends a great deal on local circumstances and personal aptitude.

Honorary membership
Since 1955, the association has appointed honorary members from amongst clinical chemists of great distinction. The list of honorary members is as follows: Dr. Donald D. van Slyke (1955-1971); Sir Rudolf Peters (1958); Dr. C. P. Stewart (1963-1971); Sir Charles Dodds (1966-1973); Professor Paul Astrup (1971); Professor Samuel A. Berson (1971); Dr. Robert Gaddie (1972); Professor Martin Rubin (1972).

Emeritus members
In 1973 it was decided to recognize great service to the association by certain of its members by electing them to the title of emeritus member. To date there are three such members, namely, H. Varley, G. Higgins, N. F. Maclagan. All of these were appointed in 1973.

UNITED STATES OF AMERICA

History
Prior to the early part of the twentieth century laboratory examinations common to the time included urinalysis, blood counts, gastric analysis, sputum examinations, and a few simple bacteriological procedures (1). In the second decade of the century the introduction of quantitative

blood and urine analysis introduced a new era of the application of chemical science in the support of medical practice. The procedures of Folin and Denis for analysis of urea, non-protein nitrogen, ammonia and uric acid in blood reported in 1912 (2) were shortly followed by methods for the determination of sugar in blood by Shaffer, Lewis *et al.* (3, 4, 5). The availability of the photoelectric colorimeter to supplement and supplant visual colorimetry in the 1920s provided sharply increased sensitivity for quantitative analysis thus decreasing the size of the biological sample required for examination. This ushered in an explosive era of development which continues at an exponential rate.

Pathologists who had become more and more involved in clinical laboratory methods organized the American Society of Clinical Pathologists in 1922. Assistants, called 'laboratory technicians' or 'clinical laboratory technicians' were educated and trained. Instruction was standardized and recognition given to qualified individuals by the formation of the Registry of Medical Technologists in 1928. By 1949 a Board of Schools, established by the Council on Medical Education of the American Medical Association and the pathologists organization was able to establish qualifications for training programs and various levels of clinical laboratory testing. These efforts have continued to the present time.

Prior to 1950, with the exception of a handful of locations, clinical laboratories were under the direction of a physician, almost invariably a pathologist. The director usually personally supervized the activities in chemistry, hematology, blood banking and microbiology. Laboratories outside hospitals were also, for the most part, directed by physician pathologists. A few independent laboratories were operated by non-medical biologists and chemists in those jurisdictions where this was permissible. The combination of the increased technical complexity of the procedures in use in the clinical chemistry laboratory and the growing recognition of the potential contribution to medicine of this field induced a growing number of pathologists to direct their talents to specialization in clinical chemistry.

Likewise professional chemists were attracted to the medical environment of the clinical chemistry laboratory and the challenge of problems in human biochemistry associated with this work. The influx of chemists in this activity has accelerated and continued to this time. The process was speeded by the formation of an organization in 1949, The American Association of Clinical Chemists, devoted entirely to the speciality of clinical chemistry. In 1950 the American Board of Clinical Chemistry was established to provide a means of certifying specialists in clinical chemistry by conferring 'diplomate' status for the highest level of professional competence. The National Registry in Clinical Chemistry, established in 1961, certifies individuals for technologist and clinical chemist competence. Public and legal recognition of the speciality of clinical chemistry came during the period 1960-1970 when various state governments and the federal government enacted legislation recognizing qualified physicians and chemists as laboratory directors and supervisors. The identification of clinical chemistry as a speciality was furthered by the publication of the journal *Clinical Chemistry* which started in 1955 and which had been initiated in 1949 as a newsletter titled *The Clinical Chemist.*

Legal status

In the United States of America legal control of clinical laboratories is vested in the 50 state governments and in the federal government for specimens which either cross state boundaries or for which analytical services are paid using federal funds.

A legal basis for the function of clinical chemistry laboratories was provided during the 1950s in California and was extended in the 1960s with the adoption of state licensing laws by New York and Illinois. By 1968 the total of states having some form of regulatory controls for clinical laboratories had risen to 19. The first significant federal development occurred in 1964 when the Health Insurance for the Aged Act, more commonly known as Medicare, was enacted. In order to determine which facilities were competent to render diagnostic service to the elderly, it became necessary for the government to enunciate pertinent conditions and qualifications for both laboratories and practitioners in the field. These conditions, which underwent several variations before their adoption in December, 1966, clearly recognized that clinical chemists were competent to direct clinical laboratories, as well as to function as technical supervisors and technologists. Heretofore, arguments had been successfully waged in support of the principle that only physicians, specifically clinical pathologists were qualified to direct clinical laboratories.

In 1967, the Congress followed up with still more significant legislation when it enacted the Partnership for Health Amendments of 1967, one section of which provided for licensing of clinical laboratories engaged in interstate commerce. Chemists were prominently involved in the consideration of this legislation through their professional societies, principally the American Association of Clinical Chemists and the American Chemical Society.

From this Act also have flowed regulations to be administered by the federal government for its enforcement. In a large part, these are quite similar to the conditions established for independent laboratories participating in the Medicare program in that they also recognize the competence of clinical chemists to serve as directors and in other supervisory and technical capabilities in such laboratories. Another major consideration is that the government, in collaboration with the scientific and medical communities is moving rapidly into the area of performance evaluation and proficiency testing of such laboratories, so that standards of service rendered the public are apt to improve notably in coming years.

Education and training for directors and senior staff of chemistry laboratories

Since present legal requirements in this country at the federal level for directors require either a medical degree or the PhD degree it is appropriate to consider this matter in detail in subsequent sections. In general it may be stated that for those with a medical degree it is required that the usual education be extended to the level of approval by the appropriate training leading to board certification in pathology. For non-medical graduates the advanced postdoctoral programs and doctoral programs as described subsequently are required for such specialization.

Place of clinical chemistry in hospital services

Clinical chemistry services in the hospital may function as an independent section of the laboratory division under a qualified director or may be provided as an integrated portion of a single polyvalent hospital clinical laboratory. In general, the larger hospitals are organized so that clinical chemistry is a separate section while the smaller hospitals (under 400 beds) provide laboratory services in a single homogeneous division. In the latter case the laboratory division is usually headed by a pathologist. In the former situation the laboratory division as a whole is usually under the direction of a pathologist with the section of clinical chemistry controlled by a qualified director with a medical or science degree.

Place of clinical chemistry in the non-hospital services

A large fraction of clinical chemistry services in the country is provided by non-hospital private laboratories. For many years most of these laboratories have been owned and directed by pathologists. It has been estimated that there may be as many as 20,000 such laboratories. In most such organizations the clinical chemical analyses have been done by medical technologists, sometimes under the supervision of an individual with special training in clinical chemistry. Relatively few such laboratories presently engage the services of a clinical chemist at the doctorate level of education but their number is growing.

A second group of private commercial polyvalent laboratories are owned and operated by bioanalysts, usually holding the doctorate degree in one of the physical sciences. Medical technologists carry out the clinical chemistry testing along with their other duties.

In recent years there has been an increasing tendency for the amalgamation of both the types of laboratories described above into large commercial chain organizations. In such cases the section of clinical chemistry is usually under the direction of a qualified PhD director of clinical chemistry.

Place of clinical chemistry in medical education—undergraduate

In almost all of the medical schools of the USA the education of the physician in clinical chemistry during his undergraduate years occurs at several points. The preclinical years of medical education provide for an introduction to the subject during the usual first-year course in biochemistry. The extent and depth of this exposure varies with the school but in general an attempt is made in all schools to relate the more theoretical aspects of biochemistry to clinical problems during formal lectures, in laboratory periods and by clinical conferences at which patients with biochemical problems are presented and discussed. In the next stage of preclinical medical education clinical chemistry is usually presented as a part of the teaching in laboratory medicine. Lectures, and patient case presentations are a common component of these courses. In recent years laboratory exercises have tended to be brief if chemistry may be provided as part of a course in pathology or laboratory medicine. An occasional elective course in clinical chemistry is offered to medical students who desire more intensive experience in the field. In the final clinical years of medical education the use of clinical chemistry laboratory testing is developed intensively as a component of patient care on the part of the medical student in the hospital environment.

Postgraduate medical education in clinical chemistry

Further speciality education in clinical chemistry is provided as part of the training for board certification in pathology. The physician in such a program usually goes through a rotation in all the sections of clinical laboratory science of which clinical chemistry is one part. The length of time in the speciality may vary from three to six months depending on the individual curriculum. Physicians who are to specialize in clinical rather than anatomical pathology will spend considerably more time in the clinical laboratory services.

Place of clinical chemistry in non-medical education

Postdoctoral programs

In a number of universities and teaching hospitals it is possible for individuals who have earned the PhD degree in one of the physical sciences or the MD degree to continue with a two-year program in specialization in clincal chemistry. These programs are highly individualized to accommodate the varied backgrounds of the participants. Those who come from a background of medicine are usually provided with an opportunity to improve their basic science understanding by participation in graduate chemistry courses and seminars. At the same time their research skills are developed by undertaking a project in the clinical chemistry department.

The larger group of participants who enter the programs with a doctorate degree in the sciences may enlarge their understanding of a given subject by class attendance and seminar participation, as in biochemistry. They will become accustomed to medical problems by attendance at hospital meetings and patient visits. All trainees are occupied primarily with activities in the clinical chemistry laboratory. Thorough grounding in all technical procedures is given by work with the departmental staff. A knowledge of routine affairs, organization, methodology and management of the clinical chemistry laboratory is imparted in this way. Subsequent activities include the evaluation of quality control in the laboratory, design and testing of new methods and survey of the literature. Programs usually also include an opportunity to serve in a supervisory and teaching capacity.

For the most part, postdoctoral study in clinical chemistry is for a period of two years. Upon completion of study in a program approved by the American Board of Clinical Chemistry a student may be admitted to board examination. Successful completion of the examination followed by the requisite years of working experience allow for the attainment of diplomate status, recognized by federal and many state jurisdictions as qualification for serving as director of a clinical chemistry laboratory. Board-approved programs are functional at the University of Florida, University of Pennsylvania, Georgetown University, University of Maryland, State University of New York at Buffalo, Ohio State University, University of Iowa, Michael Rees Hospital, Hahnemman Medical College, and the University of Washington. Additional programs have been organized at Loyola University, Mount Sinai University, Medical College of Virginia, Baylor University, University of Oregon and Louisiana State University.

Additional postdoctoral specialization in clinical chemistry is provided at several hospital locations including Hartford Hospital, The Cleveland Clinic, The City of Hope Hospital, and the Bio Science Laboratories.

A total of 66 candidates were enrolled for the above programs in 1973. Support for these students has been provided by several mechanisms. At some universities federal training grants in clinical chemistry have provided the impetus for the effort. Support for these programs has decreased since January, 1973. In many instances the candidates are provided with employment as technologists in regular hospital staff positions. A few institutions have earmarked fellowship funds which are available for postdoctoral trainees. A number of candidates are supported as research assistants on government research contracts. The current decrease in the availability of positions for graduates of doctoral programs in the physical sciences has brought about a heightened interest in postdoctoral programs in clinical chemistry. Unfortunately sufficient support is not available to provide for the training of these promising candidates.

Doctoral programs

For many years it has been possible in a few universities to combine the

178

usual doctorate program with sufficient complementary experience in the clinical chemistry laboratory to be able, upon completion of such a program, to function as an effective senior member of the clinical laboratory staff. In the last decade the number of such programs has increased and the course content has been fairly well standardized. Both the Education Committee of the American Association of Clinical Chemists and the American Board of Clinical Chemistry have established essentially uniform criteria for such education. For the most part the requirements include a background of major specialization in chemistry at the college level, the usual number of university credit hours of graduate study distributed variably according to the individual program between analytical, physical, organic and biochemistry, clinical chemistry, mathematics and statistics, instrumentation, automation and electronic data processing; seminars, some pathology and pharmacology in a few universities, foreign language comprehension and qualification, and completion of qualifying and comprehensive examinations in chemistry and the preparation and oral defense of a research thesis. Candidates in these programs have been supported as participants in nationally funded programs, as fellows in individual universities, as research and teaching assistants, and as hospital clinical laboratory employees. Present programs include those at Ohio State University, Loyola University, University of Iowa, University of Maryland, Cleveland State University, Hahnemann Medical College, University of North Carolina, University of California (San Diego), Louisiana State University and the Mount Sinai Medical Center (Chicago). Thirty candidates were enrolled in these nine programs in 1973.

Candidates from programs approved by the American Board of Clinical Chemistry may stand for board examination. Upon successful completion of the examination and requisite experience the board will confer diplomate status.

Master's degree programs
Students who have achieved a baccalaureate degree with major specialization may continue in clinical chemistry programs leading to the master's degree. Training guidelines for such programs are established by the Committee on Education of the American Association of Clinical Chemists. Graduate programs in clinical chemistry for university graduates in medical technology have been established at several universities where other programs of specialization in medical technology may also be available.

Most master's degree programs are of two years' duration and a total of 30 graduate credits are the norm. These usually include biochemistry, instrumentation, analytical chemistry, clinical chemistry, electronic data processing frequently and clinical chemistry laboratory experience invariably is a part of the curriculum. Management and supervisory skills are engendered by work experience and formal education. A modest research or development thesis in conjunction with the work of the clinical chemistry laboratory is usually required.

University programs at the master's level in clinical chemistry have been identified at the University of Dayton, Ohio State University, University of Tennessee, University of Iowa, Hahnemann Hospital, University of Maryland, Georgia State University, Cleveland State University, University of Chicago, University of California (San Diego), University of Oregon and Louisiana State University. Many others are in process of formation.

Candidates who complete these programs of study usually qualify in supervisory positions in clinical chemistry departments. Appropriate recognition of this advanced education is given by the National Registry in Clinical Chemistry.

The opportunity for college graduates with a degree in medical technology to continue to advance their education and attain the master's degree in clinical chemistry is afforded by the University of Colorado, University of Illinois, the University of Nebraska and the University of Minnesota.

The graduates of specialist programs in clinical chemistry in university schools of medical technology may qualify for the certificate in chemistry of the Registry of Medical Technologists or the NRCC.

Financial support for students in these programs is provided by stipends, scholarships, loans, faculty appointments as teaching assistants, some traineeship grants from federal and state sources and job appointments in clinical laboratories. In 1973 twelve university programs in clinical chemistry at the master's degree level had an enrolment of 90 students. A survey of 779 medical technologists engaged in graduate study in 1965 revealed that 105 were enrolled in programs in biochemistry.

Baccalaureate programs
Specialization in clinical chemistry as a major subject is now offered for the college degree (bachelor of arts, bachelor of science) in several universities. Guidelines for these programs have been provided by the Committee on Education of the American Association of Clinical Chemists. For an appropriate program a minimum of 32 credit hours in chemistry course work is suggested to include a balanced program of studies in inorganic, analytical, biochemical and physical chemistry. Instrumentation and an opportunity for practical experience in the clinical chemistry laboratory is a requisite part of the education. It has been strongly suggested that programs in this category be structured to enable students to continue in graduate study in clinical chemistry. At the present time university programs in clinical chemistry at the college level include those at the University of Southern Florida, Youngstown College, Georgia State University, The University of the City of New York, Seattle University, Quinnipiac College, New Haven

College and the University of California. A total of 136 candidates were enrolled in these programs in 1973.

Technologists, technicians and other laboratory workers

Technologists

Most technologists working in clinical chemistry have been educated and trained in college programs organized by the American Society of Clinical Pathologists through the Board of Registry of Medical Technologists and the Council on Medical Education of the American Medical Association.

Programs for education of medical technologists provide for three years of college with approved science courses and a minimum of twelve months at a school of medical technology approved by the National Accrediting Agency for Clinical Laboratory Sciences or alternatively a college science degree plus five years of laboratory experience under a pathologist and a medical technologist (ASCP) supervisor. These programs provide fundamental science education in mathematics, physics, biology and chemistry plus a year of working experience and education in the hospital environment. In the period since the start of these programs over 50,000 medical technologists have been educated and examined for qualification in the field.

In recent years technologists specialized in clinical chemistry only have been prepared by the university programs described above.

Technicians

The program for the training of technicians called Certified Laboratory Assistants (CLA) was initiated in order to provide individuals who would be capable of working as technical assistants in clinical laboratories. Potential trainees are required to have completed a high school education. By 1971 there were 148 schools which had organized such programs with capacity for 1,300 students. The training consists of twelve months in an approved school. The training was started in 1963 by the American Society of Medical Technologists. The Board of Certified Laboratory Assistants approves schools and certifies persons as CLA upon examination or by reciprocity.

In addition to the above program the various military services have established schools for the education and training of clinical laboratory technicians. Upon completion of the program of study these students may become a Certified Laboratory Assistant by a reciprocity agreement with the board.

Possible future developments

In recent years a number of conferences have attempted to forecast the future needs for scientists and workers in clinical laboratories, including the speciality of clinical chemistry. Perceptive reports have provided a cogent summary of current thinking in the field (6-14).

Clearly the new legislation mandating requirements of education and training for specialists in clinical chemistry will sharply increase the need for qualified directors in the immediate years ahead. The hope of obtaining a supply of such individuals from the usual channels of postgraduate medical education is not optimistic. On the other hand the sharply decreasing financial support for training programs by federal funding appears to sharply inhibit the promising development of the last few years of university programs in clinical chemistry. This is especially unfortunate because the newly awakened interest of scientists in human biochemistry has focused the attention of capable young investigators upon this field. There is no question but that the requirements of health care delivery will demand improvement in the quality control, the accuracy and the performance of tests. Automation and computer correlation of laboratory data, diagnosis and therapy will be part of the expanding services of the clinical laboratory scientists. Specific enzymatic procedures, drug analysis in clinical pharmacology, endocrine assays, immunological testing, tests for fetal wellbeing as well as the health of the pregnant woman, requirements of microanalysis for pediatric patients, screening for genetic defects and the entire problem of preventative health and environmental healthfulness will demand the highest caliber of scientific skills. It is equally clear that we are on the threshold of major advances in the biochemistry of mental disease which will require the application of increasingly sophisticated laboratory performance. The role of clinical chemistry in the diagnosis and treatment of degenerative disease and cancer will surely challenge the highest scientific capabilities. In the face of these future developments the need is clear for the education and training of competent scientists in the field.

It is sometimes suggested that the advent of automated instrumentation in clinical laboratories will minimize the need for extensive technologist training. In fact, the opposite conclusion appears to be valid. Complex instrumentation demands higher rather than lesser skills for understanding and control. Education of technologists in instrumentation, electronics and computer skills has become a mandatory addition to previous training programs.

The development of regionalized specialized laboratories for clinical chemistry will bring with it an increasing demand for trained specialists. It is essential that educational program at all levels be encouraged and extended in clinical chemistry.

National societies, associations and journals concerned only with clinical chemistry

The American Association for Clinical Chemistry

The association was officially incorporated as a non-profit professional organization in the State of Delaware in 1950. It is composed of local autonomous groups in the form of sections located in various parts of the country. The administration of the association is in the hands of its officers, elected by all the members of the association. The officers are charged with the executive fulfilment of association policy which is established by a board of directors, which is elected by the vote of the members of the association.

Since its formation the association has sought to advance the practice of clinical chemistry in the USA. It has done this by holding national meetings each year in alternative areas of the country. The national meetings are marked by the presentation of formal lectures and symposia of leading authorities in the field. Individual scientific communications from the membership, demonstrations and workshops provide the balance of the scientific presentations. At the national meetings manufacturers of instruments, chemicals and books exhibit their most recent developments. Various committees of the association have been active in advancing the field. The Education Committee fosters training programs at various professional levels in the colleges and the universities. The Standards Committee has been active in developing suitable quality control materials, standards and methods for use in the clinical chemistry laboratories. The Publications Committee supervizes the journal, *Clinical Chemistry* and arranges for publication of the continuing series of *Standard Methods in Clinical Chemistry*.

The local sections also hold regional scientific and social meetings usually four or five times each year. The association presently has over 3,000 members.

The American Board of Clinical Chemistry

The American Board of Clinical Chemistry was organized in 1950 under the sponsorship of the American Chemical Society, the American Institute of Chemists, and the American Society of Biological Chemists. The American Association of Clinical Chemists became a sponsoring society in 1954. Bylaws were adopted at an organizational meeting in Atlantic City, New Jersey, on April 18th, 1950. A Certificate of Incorporation was granted by the State of Delaware on August 23rd, 1950, and the first annual meeting of the board was held in Chicago, Illinois, May 25th-27th, 1951.

The principal function of the board, in the interests of the public and the advancement of the science, is to establish and enhance standards of competence for the practice of clinical chemistry, including toxicological chemistry; and to certify as qualified specialists those voluntary applicants who satisfy the requirements of the board. In this way, the board aims to ensure to the public, both lay and professional, that persons professing to be specialists in clinical chemistry possess the requisite qualifications and competence. The action of the board is based upon the candidates ethical and professional record, education experience, and attainments, as well as the results of a formal examination.

One of the objects and purposes of the board stated in its Certificate of Incorporation is the preparation and distribution in the public interest, of a registry of persons with specialized knowledge in clinical chemistry who have been granted certificates by the board. The first such directory was published by the board in 1954. The directory is available without charge to hospitals, universities, laboratories engaged in clinical chemistry, and to others with a legitimate need for this information. With the acceptance, in principle, in 1966 by the federal government that chemists proficient in clinical chemistry could serve as directors of clinical chemistry laboratories as far as federal regulations is concerned, the status and activities of the American Board of Clinical Chemistry has entered a new phase of importance. The federal regulations have been also accepted by many state jurisdictions thus recognizing certification of the American Board as acceptable legal justification for the post of director of the clinical chemistry laboratory. As a consequence of these developments the interest of chemists in the certification program has increased markedly. Examinations are given nationally for candidates who qualify by education and experience. Educational requirements are a doctorate degree in the medical, biological or physical sciences with sufficient and appropriate training in chemistry. Experience requirements consist of some years of responsible work in an approved laboratory of clinical chemistry. When qualified by the above criteria admission to examination has been made for those students who have completed one of the board-approved training programs in clinical chemistry. Such a candidate may be admitted to examination before completion of the work experience requirement. Upon satisfactory completion of the examination and when work experience requirements have been satisfied the candidate is admitted to diplomate status of the American Board of Clinical Chemistry. Somewhat under 700 candidates have been admitted since the start of the American Board in 1950.

The following clinical chemistry training programs were approved by the board as of March, 1976: Buffalo General Hospital, Cleveland State University, Georgetown University School of Medicine, Hahnemann Medical College and Hospital, Hartford Hospital, Louisiana State University, Loyola University, Stritch School of Medicine, Medical College of Pennsylvania, Michael Reese Medical Center, National Institutes of Health, State University of New York at Buffalo, The Ohio State University, University of Iowa, University of North Carolina, School of Medicine, University of Oregon Medical School, University of Washington and University of Windsor.

The National Registry in Clinical Chemistry
This was created by sponsorship of:

> The American Association of Clinical Chemists
> The American Board of Clinical Chemistry

The American Chemical Society
The American Institute of Chemists
The American Society of Biological Chemists

to identify qualified chemists and technologists at the baccalaureate level who provide essential public health services in the nation's clinical laboratories. As a reflection of views held by the sponsoring organizations and in line with present federal standards for clinical laboratories, the registry grants accreditation at two levels, namely, 'clinical chemistry technologist' and 'clinical chemist'. The category of 'clinical chemistry technologist' is designed primarily for applicants with recent bachelor's or master's degrees in chemistry or for those with academic degrees in other disciplines who regularly perform clinical chemistry determinations.

The registry annually compiles and publishes a list of all individuals accredited during the year who meet the following standards:

[a] *Clinical chemistry technologist*—Applicants for accreditation as a clinical chemistry technologist must possess a minimum of a bachelor's degree with a major speciality in one of the chemical, physical, or biological sciences from an institution acceptable to the registry, including at least 16 semester hours (or equivalent) of appropriate college level studies in chemistry.

Applicants also must have acquired a minimum of one year of acceptable experience in clinical chemistry subsequent to attaining the bachelor's degree.

[b] *Clinical chemist*—Applicants for accreditation as a clinical chemist must possess a minimum of a bachelor's degree in chemical science or in a closely related discipline, in either case from an institution acceptable to the registry, including at least 32 semester hours (or equivalent) of appropriate college level studies in chemistry.

Applicants also must have acquired a minimum of six years of acceptable experience in clinical chemistry subsequent to attaining the bachelor's degree. Appropriate graduate education may be substituted for the required experience on the following basis:—master's degree: two years; earned doctor's degree: four years.

Societies concerned with all clinical laboratory specialities

In the United States of America, as in many other countries, a number of organizations are concerned with all components of clinical laboratory services. Many of these have not only pioneered the development of clinical chemistry but continued to make major contributions to the advancement of this speciality.

The College of American Pathologists and the American Society of Clinical Pathologists

These organizations, concerned with anatomical and morphological pathology and with all aspects of clinical laboratory activities respectively, have provided more than half a century of leadership in education, research, initiation of quality control programs, workshops, seminars and regional and national meetings devoted to all facets of the subjects. They have been concerned with matters of public policy, with legislation, laboratory management and the economics of the subject. Space will not permit the presentation of detailed information of their many activities, which can be obtained directly from the organizations.

The American Society of Medical Technologists
This society was organized originally as the American Society of Clinical Laboratory Technicians in 1933. Three years later when the distinction was made between 'technologist' and the less well-educated and trained 'technician', the name of the society was changed. The national organization and its constituent societies provide continuing education through academic courses, workshops, annual conventions, seminars, in-service training programs and postgraduate activities. The membership has doubled in the last decade to 16,000 members requiring an administrative staff of 26 full-time employees. A Scientific Assembly divided into ten scientific sections, one of which is biochemistry, develops the speciality interests of the members. The society publishes a monthly *Newsletter, The American Journal of Medical Technology* and a bimonthly publication entitled *'Cadence in the Clinical Laboratory'* which is a non-technical presentation of laboratory problems.

The American Association of Bioanalysts
This was formed in 1965 from a merger between the Council of American Bioanalysts and the National Association of Clinical Laboratories. The association represents the scientific and professional interests of scientists who have expertise in the various laboratory disciplines and who have achieved the ability to serve as directors of the general bioanalytical laboratories. Certification and accreditation of such individuals is carried out by the independent American Board of Bioanalysis upon successful completion of examination for those who meet the requirements of elegibility for examination. Two states presently recognize and licence bioanalysts as laboratory directors. In addition to sponsoring the board, the association has a Division of Professional Services which provides an external laboratory evaluation service. The division also maintains a department of Continuing Education which conducts workshop programs in a number of universities in the State of California. The association publishes a bulletin concerned with bioanalytical laboratory developments.

The American Medical Association Council on Medical Education
In collaboration with the National Accrediting Agency for Clinical Laboratory Sciences this serves as an accrediting body for programs in

the allied medical services. By 1963 the council had accredited 30 schools and colleges and 759 hospitals offering approved programs in medical technology.

The Board of Registry of Medical Technologists of the American Society of Clinical Pathologists
This board was established in 1928 to standardize the training of medical technologists. The board is a standing committee of the American Society of Clinical Pathologists and includes five members of the American Society of Medical Technologists. By review of applications and by examination the board provides certification of individuals as a medical technologist and also as a specialist by award of a certificate in chemistry. Its function has now been assumed by the National Accrediting Agency for Clinical Laboratory Sciences.

The Board of Certified Laboratory Assistants of the American Society of Clinical Pathologists
This is composed of four members of the society and four members of the American Society of Medical Technologists. By examination of individuals from approved schools, by reciprocity with approved military programs or with the respective societies, the board may provide certification.

The National Committee for Careers in the Medical Laboratory
The committee is a non-profit organization concerned primarily with the recruitment and education of clinical laboratory manpower. It has sponsored many special projects designed to increase the numbers and improve the quality of education programs for medical laboratory workers, including demonstration programs, surveys, training manuals, teacher training institutes, and educational and recruitment films and publications.

The committee was organized by and is sponsored by the American Society of Clinical Pathologists, the College of American Pathologists and the American Society of Medical Technologists.

Academic Clinical Laboratory Physicians and Scientists
This organization, composed of educators in the field of clinical laboratory science is concerned with problems in education as well as clinical laboratory practice. It sponsors national meetings on the subject.

Association of Clinical Scientists
The association is a national organization concerned with clinical laboratory science in all its ramifications. Membership is open to physicians and other scientists concerned with this field. National meetings are held twice a year. Lectures, seminars, workshops and demonstrations are presented. The association maintains a proficiency

testing service which provides for external laboratory control. The association was established in 1949 and has approximately 350 members.

Journals devoted solely to clinical chemistry
'Clinical Chemistry', the journal of the American Association of Clinical Chemists, commenced publication with Volume I in 1952. The association also publishes the serial volumes of *'Standard Methods in Clinical Chemistry'*.

REFERENCES

1   Osler, W. (1898) *The Principles and practice of medicine*, D. Appleton and Company, New York.

2   Folin, O., and Denis, W. (1914) New methods for the determination of total non-protein nitrogen, urea and ammonia in blood, *J.Biol.Chem.*, 11, 527.

3   Shaffer, P. A. (1914) On the determination of sugar in blood *J.Biol. Chem.* 19, 285

4   Lewis, R. C., and Benedict, S. R. (1920) A method for the estimation of sugar in small quantities of blood, *J.Biol.Chem.* 20, 61.

5   Myers, V. C. (1921) *Practical chemical analysis of blood*, C. B. Mosby Co., St. Louis, Missouri, pps. 65-68.

6   Kinney, T. D., and Melville, R. S. (1969) The clinical laboratory scientist, *Laboratory Invest.* 20, 382.

7   Grumer, H. (1970) The training of clinical chemists, *J.Chem.Education* 47, 765

8   Jones, B. (1967) Education for the allied health professions, US Dept. of HEW, Public Health Serv., Publication 1600, US Government Printing Office, Washington, DC, 20402.

9   Coon, R. (1967) *Manpower for the medical laboratory, US* Government Printing Office, Washington, DC, 20402

10   Kinney, T. D., and Melville, R. S. (1967) Automation in clinical laboratories, *Laboratory Invest.* 16, 803.

11   Baver H. (1967) The national laboratory crisis *Hospital Practice*, p.1.

12   Straumfjord, J. V. (1967) Report of ad hoc committee on reorganisation of laboratories, *Bulletin College of American Pathologists*, XXI 108.

13   Sturm, H. M. (1967) Technological developments and their effects upon health manpower, United States Department of Labor, *Monthly Labor Review*, Reprint No. 2516.

14   Smith, C. E. (1967) Educational qualifications of public health laboratory workers, *American Journal of Public Health*, 57, 523.

VENEZUELA

History

For the last 19 years the term 'bioanalysis' has been utilized to correspond with 'clinical chemistry' in other countries.

In 1938 there were very few laboratory technicians. The very scattered teaching in this field was provided in private schools. The course was from six months to three years. In 1949 the School of Medicine of the Central University of Venezuela (in Caracas) created the School for Clinical Laboratory-Technicians. It was a two-year course. In 1950 the University of the Andes (in Mérida) created The Politechnical School for Laboratory-Technicians as a part of the School of Pharmacy. This was a three-year course. In 1956 there was unification in Venezuela and The School of Bioanalysis was founded in the University of the Andes as a three-year course.

In 1962 following completion of the requirements of the National Law for Universities, a four-year course was established as the basic preparation for this profession and the degree of 'licentiate in bioanalysis' was granted. In 1968 the University of Zulia (in Maracaibo) created The School of Bioanalysis. In 1970 upon completion of the legal requirements for the presentation and defense of a doctoral thesis, it became possible to obtain the degree of 'doctor in bioanalysis'. In 1972 the National Council of Universities approved the creation of The School of Bioanalysis in the University of Carabobo as a five-year course.

Legal status

In 1974 the National Congress approved the Law for the practice of bioanalysis, but up to the present time, curriculum requirements have not been approved. Each local college has its own statutes.

The director of any laboratory should have the degree of 'bioanalyst' (laboratory technician). A medical doctor can be the director of a laboratory if he has specialized in clinical pathology or has had three years of specialization in a particular branch of bioanalysis.

Societies

In each state there is a local 'college' which unifies the specialists of this profession. This 'college' then constitutes what is known as The National Federation of Colleges of Bioanalysis in Venezuela (founded in 1960). The 'college' mentioned here is to be understood as the local organization, meeting place, etc., for a group of professional people in a determined area of the country. The members of this 'college' may be:

> Doctors in bioanalysis
> Licentiates in bioanalysis
> Bioanalysts
> Laboratory technicians

189

Clinical chemistry in the university

There are four Schools of Bioanalysis in Venezuela:

(1)   The School of Bioanalysis of The University of Carabobo (in Valencia), which forms part of the School of Medicine.

(2)   The School of Bioanalysis of the Central University of Venezuela (in Caracas), which forms part of the School of Pharmacy.

(3)   The School of Bioanalysis of the University of the Andes (in Mérida), which forms part of the School of Pharmacy.

(4)   The School of Bioanalysis of the University of Zulia (in Maracaibo), which forms part of the School of Medicine.

The plan of study in these schools includes one year of basic study and three years of professional studies, which lead to the degree of 'licentiate in bioanalysis'. Upon the presentation of a doctoral thesis, the degree of doctor in bioanalysis is granted. There are various hospitals which contribute to the university instruction in their departments of bacteriology, mycology, hematology, parasitology and biochemistry. Good libraries are also available which are in the process of enlargement and improvement.

Laboratories

There are three types of laboratories:

(1)   Laboratories which depend upon the Ministry of Health and Social Assistance.

(2)   Laboratories which depend upon the different universities.

(3)   Private laboratories.

YUGOSLAVIA

Legal status

Federal health authorities regulate the medicobiochemistry laboratories in Yugoslavia. Two basic groups of professionals are qualified for work in the field of medical biochemistry: highly qualified professionals with academic degrees, and laboratory technicians with medium or advanced training. The former group comprises:

(1)   Medical biochemists
(2)   Specialists in medical biochemistry
(3)   Masters of science in medical biochemistry
(4)   Doctors of science in medical biochemistry

The first two groups have a professional degree, the latter a scientific degree. The latter group includes junior and senior laboratory technicians.

Education and training for directors and senior staff of clinical chemistry laboratories

The basic training requirement for medical biochemists is graduation from one of the following university schools: faculty of pharmacy, medical school or faculty of chemistry.

Medical biochemists

Most medical biochemists in Yugoslavia are graduated from the faculty of pharmacy, probably because the teaching program of these faculties in Zagreb and Belgrade special emphasis is laid on subjects relevant to training in medical biochemistry. Zagreb University has recently set up a pharmaceutical-biochemical faculty at which regular courses ramify in three directions, one of them being medical bio-chemistry. Throughout the latter course, as is clearly seen in the attached curriculum (Annex 2) teaching is specifically designed to train biochemists. After completion of these courses the student obtains a title of medico-biochemical engineer.

The title of medical biochemist is conferred upon a medicobiochemical engineer after his undergraduate training and upon pharmacists, physicians, and chemists after a postgraduate training in medical biochemistry and successful final examination. This qualifies them for work in medicobiochemical laboratories, and also for a superior position in less-developed laboratories supervized by a specialist in medical biochemistry.

Specialist in medical biochemistry

A medical biochemist, after passing the state specialist examination in medical biochemistry, obtains the title of specialist in medical bio-chemistry and is qualified for heading a biochemical laboratory and supervizing the work of less-developed laboratories of this kind.

Master of Science in medical biochemistry

A medical biochemist, after completing postgraduate training in medical biochemistry at either the medical, or pharmaceutico-biochemical faculty and successfully maintaining his thesis, obtains the scientific degree of master of science in medical biochemistry. In addition to carrying out medicobiochemical work these medical biochemists usually take part in teaching and research.

Doctor of Science in medical biochemistry

When a medical biochemist with a master's degree is successful in his research work, he may prepare a thesis for the doctor of science degree and, if successful, obtain the scientific degree of doctor in medical biochemistry.

Specialist education programs

There are two ways to qualify for the professional title of medical biochemist: firstly to complete the respective university study, or secondly, to go in for postgraduate training.

Medical biochemist

At the medicobiochemical department of the pharmaceutical-bio-chemical faculty, the teaching is arranged in such a way as to qualify the graduates, i.e. medicobiochemical engineers, to obtain the title of medical biochemist (see Annex 2).

The title of medical biochemist through postgraduate training is open to graduates in pharmacy, medicine, and chemistry. Additional post-graduate training in medical biochemistry, lasting two to four semesters, is organized at the medical school and the pharmaceutical-biochemical faculty.

The postgraduate training is carried out according to the attached curriculum (Annex 3). Special attention is paid to practical training. Thus, every morning the candidates are included in the regular work of the medicobiochemical laboratory designed as a training basis. They carry out analyses, in groups of two to three, and are supervized by a specialist in medical biochemistry. In the afternoon they attend regular lectures and seminars. Out of the total number of teaching hours, two-thirds relate to chemistry and only one-third to all other allied subjects. After completing the study, the candidates take the final examination before a board composed of lecturers having taken part in the respective postgraduate course. The diploma is issued by the faculty at which the study is organized. If the candidate, after passing the final examination, prepares a dissertation for the master of science degree and maintains it before a board composed of respective professors, he obtains the scientific degree of master of sciences in medical biochemistry.

Specialist in medical biochemistry

The medical biochemist, after working in a medicobiochemical labora-tory for three years, is entitled, in agreement with the institution in which he works, to apply to the Secretariat of Public Health and Social Welfare for specialization in medical biochemistry. The Secretariat gives him permission to start his specialist internship and selects an institu-tion for it. Preference is given to candidates with higher average marks during their studies and success during their three-years work in medi-cobiochemical laboratories. After four months of internship the candi-date takes an examination before a board composed of the institution's experts. If he passes the examination, he continues his internship but if he fails in it, his specialization is discontinued and both the Secretariat and the institution that sent the candidate to specialize are advised about it.

Specialization lasts for three years. The first two years are spent in the medicobiochemical laboratories of university hospitals, and other health institutions selected by the Secretariat. Only laboratories headed by or manned with specialists in medical biochemistry can be considered for the purpose. The third year of specialization is spent at the faculties with organized postgraduate training in medical biochemistry. At the end of the first year of specialization the candidate takes a practical oral examination in hematology, bacteriology, and parasitology before a board appointed by the Secretariat at the suggestion of the faculty by which the postgraduate training is conducted. At the end of the second year there is an examination, both practical and theoretical, with particular reference to medicobiochemical analyses. The board of examiners is the same as that appointed for the examination following the first year of specialization. At the beginning of the third year, in agreement with one of the teachers lecturing in the postgraduate course, the candidate selects a topic for his state specialist examination paper. The paper is meant to be completed in the course of that year. If there are some serious reasons, its completion may be postponed an additional year at the most. After passing all examinations and preparing his written work, the candidate takes the state specialist examination.

Annex 1.   Faculty of pharmacy
The program is as follows:

First and second semester
General and inorganic chemistry
Applied chemistry
Qualitative analytical chemistry
Mathematics
Organic chemistry
Pharmaceutical botany with biology
Stoichiometry

Third and fourth semester
Organic chemistry
Physical chemistry
Quantitative analytical chemistry
Physiology with anatomy
Medical microbiology
General biochemistry

Fifth and sixth semester
Pharmacognosy I
Pharmacognosy II
Pharmaceutical chemistry I
Pharmaceutical chemistry II
Biological drugs
Sterilization

Pharmacology
History of pharmacy
Pharmaceutical technology I

Seventh and eighth semester
Pharmaceutical technology II
Clinical chemistry
Hematology
Analysis of foodstuffs
Parasitology
Pathophysiology
Fundamentals of economics and organization of
    pharmaceutical service
Introduction to research work (seminars)

**Annex 2.   Pharmaceutical-biochemical faculty, medicobiochemical
department**

First and second semester
Physics
Higher mathematics
General and inorganic chemistry
Stoichiometry
Qualitative analytical chemistry
Quantitative analytical chemistry

Third and fourth semester
Analytical chemistry, instrumental analysis
Organic chemistry
Biochemistry
Physical chemistry
Botany with the fundamentals of general biology
Anatomy and physiology of Man
Microbiology
Physical education

Fifth, six, seventh and eighth semester
Medical biochemistry
Biochemistry II
Hematology with the fundamentals of immunohematology
Microbiology with the fundamentals of serology
Biochemistry of foodstuffs
Pathophysiology
Dynamic biochemistry of drugs
Parasitology
Selected methods of medical biochemistry
Selected chapters of physiology

Fundamentals of statistics
Fundamentals of social sciences
Detection and protection of radiation
Fundamentals of war medicine
Written work of the diploma

Annex 3.   Postgraduate course in medical biochemistry
(Andrija Stampar School of Public Health, Medical Faculty,
University of Zagreb).

First semester
Methodology of research work statistics
Medical biochemistry
Physical methods in biochemistry
Dynamic biochemistry
Hematology
Pathohistological techniques
Clinical virology

Second semester
Hematology
Medical biochemistry
Physical methods in biochemistry
Dynamic biochemistry
Clinical parasitology and mycology
Laboratory basis for clinical diagnosis
Fundamentals of cytodiagnostics

Training of technologists and technicians
There is a regular training of medium and higher grade laboratory
technicians at the School of Health Technicians. The school has four
branches, one of them relating to laboratory work. Admission requirements for the latter is the primary school (eight years) completed with
very good marks and a successfully passed admission examination
taken before a board of examiners of the schools of teachers. The
training lasts four years (Annex 4). Since special attention is paid to
practical work throughout this schooling, the third and fourth year
trainees are included for three months in routine morning work of one
of the medicobiochemical laboratories collaborating with the school.
They carry out this work under the supervision of a senior technician.
After four years, the trainees take the final examination before a board
consisting of the schools teachers. The examination includes a written,
a practical, and a theoretical part. Only those obtaining a positive mark
for their written part are allowed to sit for the practical and theoretical
part of the examination. The trainees that pass the examination obtain
a certificate which gives them the title of junior laboratory technician.

Junior laboratory technicians, successful in their work, may be recommended by their institution for admission to the higher school for health technicians, department for laboratory work. Only those among them passing the admission examination before a board consisting of the school's teachers are admitted. The training lasts two years (Annex 5). In the second year of training the trainees are included in the regular morning routine work of one of the medicobiochemical laboratories collaborating with the school. After completing the second year training the trainees take individual examinations for the final examination, before the school's board, in the following subjects: medical bio-chemistry, hematology, microbiology, and pathophysiology. The final examination comprises a written, a practical, and a theoretical part. Only the candidates with positively marked written work are allowed to take the practical and the theoretical part. After passing the final examination, the candidates are given a diploma and the title of senior laboratory technician.

Annex 4. School for health technicians

The following subjects are taught in a four-year program: literature and fundamentals of linguistic and aesthetics, foreign language, Latin language, mathematics and health statistics, physics, chemistry, medical biochemistry, biology, anatomy and physiology, public health administration, health and social legislation, hygiene with the fundamentals of epidemiology and disinfection, microbiology and parasitology, pathophysiology and fundamentals of histological technique, laboratory analysis of water foodstuffs, hematology, physical education.

Annex 5. Higher school for health technicians

The following courses are taught in a two-year program: hygiene and social medicine, sociology, statistics, fundamentals of economics and financing of health services, foreign language, physiology with pathophysiology, chemistry, microbiology with the diagnostics of infectious diseases, clinical parasitology, histological pathology and techniques, psychology.